Managing the Modular Course

Other titles recently published under the SRHE/Open University Press imprint:

Michael Allen: *The Goals of Universities*
William Birch: *The Challenge to Higher Education*
David Boud *et al*: *Teaching in Laboratories*
Heather Eggins: *Restructuring Higher Education*
Colin Evans: *Language People*
Derek Gardiner: *The Anatomy of Supervision*
Gunnar Handal and Per Lauvås: *Promoting Reflective Teaching*
Vivien Hodgson *et al*: *Beyond Distance Teaching, Towards Open Learning*
Peter Linklater: *Education and the World of Work*
Graeme Moodie: *Standards and Criteria in Higher Education*
John Pratt and Suzanne Silverman: *Responding to Constraint*
Majorie Reeves: *The Crisis in Higher Education*
John T. E. Richardson *et al*: *Student Learning*
Derek Robbins: *The Rise of Independent Study*
Gordon Taylor *et al*: *Literacy by Degrees*
Malcolm Tight: *Academic Freedom and Responsibility*
Susan Warner Weil and Ian McGeil: *Making Sense of Experiential Learning*
Alan Woodley *et al*: *Choosing to Learn*

Managing the Modular Course

Perspectives from Oxford Polytechnic

David Watson
with John Brooks, Chris Coghill, Roger Lindsay and David Scurry

The Society for Research into Higher Education
& Open University Press

Published by SRHE and
Open University Press
12 Cofferidge Close
Stony Stratford
Milton Keynes MK11 1BY, England

and
242 Cherry Street
Philadelphia, PA 19106, USA

First Published 1989

British Library Cataloguing in Publication Data

Watson, David
Managing the Modular Course: perspectives from Oxford Polytechnic.
1. Great Britain. Higher education institutions. Modular degree courses.
I. Title
378'.1552

ISBN 0–335–09561–5
ISBN 0–335–09560–7

Library of Congress Catalog number available

Typeset by Scarborough Typesetting Services
Printed in Great Britain by
St Edmundsbury Press, Bury St Edmunds

For our students on the Modular Course –
past and present

Contents

List of Figures and Tables

Figures

Tables

Notes on Contributors

David Watson has been Dean of the Modular Course at Oxford Polytechnic since 1981. Before this he was for six years programme leader for the CNAA degrees in humanities at Crewe and Alsager College of Higher Education. His academic interests are in the history of ideas, and higher education policy, and he is the author of a number of studies of British and American ideas including *Why is there no Socialism in the United States?* (1980) and *Margaret Fuller: An American Romantic* (1988). He has served on boards and committees of the CNAA since 1977 and is currently Chair of the Committee for Institutional Relationships. He is a founder member of the PCFC.

John Brooks is a historian, with twenty years' teaching experience in higher education. He has published work on the local history of what was formerly North Berkshire. He has taught on the Modular Course for twelve years and is currently the Course Co-ordinator. He served what has proved to be a valuable 'apprenticeship' for this position by acting as senior tutor for humanities from 1978 to 1982 and history Field Chair from 1982 to 1986.

Chris Coghill, current Deputy Head of Computer Services at Oxford Polytechnic, is particularly involved with the provision of computing services for both specialist and non-specialist staff and student users, across the broad range of the Polytechnic's courses. Since 1971 he has worked in university and polytechnic administration, in finance, registry and personnel offices. For most of that time he has been both a user and a provider of computing services. For eight years, until 1987, he was both Modular Course Administrator and Systems Officer, the latter role involving the development of integrated information systems for the administration and management of the Polytechnic.

Roger Lindsay began his research career as a philosopher, but switched to psychology because he wished to know whether his beliefs were true. His doctoral research was concerned with natural language understanding, which remains a major research interest. He is keen to assist in developing psychology as an applied discipline and his work in evaluation is one part of the effort. Dr Lindsay has worked within the Modular Course since 1974 and has been Field Chair (twice), senior tutor for social studies, and Modular Course Evaluation Officer. His permanent post is as Principal Lecturer in psychology.

Dave Scurry is a geologist, appointed to the Polytechnic as a lecturer in 1974, a year after the start of the Modular Course. In 1975 he became senior tutor for science, admitting students to the fields of geology, physical sciences, mathematics, computing and cartography. In 1981 he was appointed Modular Course Co-ordinator, initially as a secondment, then permanently. As Co-ordinator he had overall responsibility for admission and student counselling. In 1987 he was appointed Assistant Dean with responsibility to the Dean for the day-to-day management of the Course. He is currently Chief Examiner for A level geology for the London Board.

Foreword

The current enthusiasm for modular course developments in both the public and university sectors of higher education comes as no surprise to the Council for National Academic Awards. From the early 1970s onwards several institutions with courses validated by the Council saw the adoption and development of such schemes as the one way forward in attempting to achieve desirable objectives such as interdisciplinary cooperation, more flexible approaches to entry into higher education, and ease of student transfer between institutions. Validation of these schemes sometimes caused anxiety and tension among the academic establishment but many misgivings were overcome as the pioneering schemes proved that they could bring about flexibility while retaining quality.

In 1989, new challenges face polytechnics and colleges particularly as most of them become higher education corporations. Their courses not only have to demonstrate efficiency in the deployment of resources but also effectiveness in their styles of teaching and learning. To the demands of the 'traditional' school leaver for relevant and responsive courses have been added those of 'new' students (part-time, mature and mid-career).

The logic of development points to modularity and credit accumulation, and the institutions which have adopted such policies have gained a deserved reputation for responsiveness.

Against this background, David Watson and his colleagues at Oxford Polytechnic have performed an important task in demystifying the modular course. The story of the Oxford Polytechnic Modular Course is one of significant and sustained success, but the authors have been frank about the difficulties which the Polytechnic faced in developing in this way and some which remain. They also perform a service in teasing out the principles of modular course organization which can operate independently of the Oxford system. The Polytechnic's achievement in initiating and developing its Modular Course over the last fifteen years is considerable, and this record should give food for thought to teachers and administrators throughout higher education.

Dr Malcolm Frazer
Chief Executive
Council for National Academic Awards

Acknowledgements

Managing the Modular Course is, as we hope is clear from the account in this book, above all an exercise in collaboration. Consequently any list of individuals who have contributed to policy, systems, and even that elusive thing, management 'style', runs the risk of becoming invidious. None the less the authors of these chapters agree that mention should be made of several colleagues and former colleagues who have played particular roles in generating the material in this book. Special thanks are due to Mark Bannister, Cheryl Codling, Brian Clark, Graham Gibbs, Martin Haigh, John Isaac, Nancy Jenkins, Mike Picken, Stuart Mossop, Robert Murray, David Turner and John Westcott. We would also like to record here our admiration and gratitude for the pioneering work on the Oxford Polytechnic Modular Course of David Mobbs, its first Dean.

David Watson
Oxford, 1989

The Oxford Polytechnic Modular Course

The Modular Course: a Brief Guide

The Modular Course has been designed to allow students to choose their own programme of study by selection from among 700 'modules' (course units) available. Each module represents between one-third and one-quarter of a term's work for most full-time students. A student progresses through the course by accumulating credits for each module passed. A certificate can be achieved by studying 10 modules; a Diploma in Higher Education by studying 20 modules; a degree by studying 28 modules; and an honours degree by studying 30 modules.

Most Modular Course students combine the study of two 'Fields' – a Field being roughly equivalent to a single subject like law or psychology. Others study a Double Field which concentrates on a subject area such as earth sciences or human biology. In all cases, students on the Modular Course follow a course of study which is academically coherent, but still allows freedom to pursue interests from a wide range of subjects.

It is by far the largest scheme, in terms of the choices available, in British higher education. In terms of the numbers of students taking part, it is by far the largest scheme outside the Open University.

(From the Polytechnic *Full-time Prospectus for 1989 Entry*)

Glossary of terms

The modules

A *single module* is a unit of study, usually lasting one term, in which one credit is earned.

A *double module* is one for which the academic content is twice that of a single module and for which twice the credit of a single module is given.

A *basic module* is one normally taken in the initial part of the course (Stage I), and represents approximately one-twelfth of the effort of a full-time honours candidate (including private study) in one year.

An *advanced module* is one normally taken in the final part of the course (Stage II), and represents approximately one-tenth of the effort of a full-time honours candidate (including private study) in one year.

A *further module* is one that a student takes in Stage II.

An *acceptable module* is one that counts towards the minimum number required for a particular field, in addition to the compulsory basic modules.

A *compulsory module* is one that must be taken and passed.

A *recommended module* is one which the student is advised to take.

A *prerequisite module* is one which must be taken before a subsequent specified module.

An *unattached module* is one that does not form part of any field.

A *synoptic module* is compulsory for honours in some fields instead of a project or dissertation module.

Other key terms

A *field* is a group of educationally associated modules, normally including 3 compulsory basic and 15 or more acceptable advanced module credits.

A *double field* is a group of educationally associated modules normally equivalent to two single fields.

A *programme* consists of the modules taken by a student over one or more terms.

A *course* is a set of programmes leading to a particular qualification.

Stage I is the initial part of the course, normally consisting of basic modules which must be completed before entering Stage II.

Stage II is the final part of the course and consists mainly of advanced modules.

A *project/dissertation* is a written report of an investigation by a student of a suitable academic topic.

A *synoptic examination* is compulsory for some fields and is taken at the end of Stage II.

(From the *Modular Course Handbook*, September 1987)

List of acronyms

ASC	Academic Standards Committee
CATE	Council for Accreditation of Teacher Education
CNAA	Council for National Academic Awards
CRCH	Central Register and Clearing House (for applications to BEd courses)

FRG	Faculty Review Group
FTE	Full-time equivalent
IFS	Interfaculty Studies
MAC	Modular Admissions Committee
MCC	Modular Course Committee
MEC	Modular Examinations Committee
MMRC	Modular Management and Review Committee
NAB	National Advisory Body for Public Sector Higher Education
PCAS	Polytechnics Central Admissions System
PCFC	Polytechnics and Colleges Funding Council
PSHE	Public Sector Higher Education
SEEC	South East England Consortium for Credit Transfer

1

Introduction: The Container Revolution

David Watson

'Modularity' is perhaps *the* buzz word of secondary and higher education in the 1980s. A *Guardian* cartoon in September 1987 shows a discomfited candidate for a job confronting an appointment panel which is dissolving into laughter. A friendly member of the panel finally leans forward and says 'Try to ignore them – it's just that they had a bet on that you'd say "modular approach" at least twice in the first two minutes'.

As usual the best humour combines self-recognition and unease. From the mid-1970s onwards school and college level curriculum reforms calling themselves 'modular' have been met with enthusiasm by a committed minority and distrust by a concerned majority of teachers at both levels. In the 1980s, with increased sensitivity to the partnership between education and the world of work and the possibility of shared learning, the notion of breaking up the educational experience into discrete blocks ('units', 'credits' or 'modules') has gained further currency.

In the public sector of higher education Oxford Polytechnic was one of the first in the field. Its undergraduate credit accumulation scheme – The Modular Course – began in September 1973 as a course in science, replacing the London University External Degree, enrolling 75 students on 7 'fields' of study. In 1988–89 it is catering for the needs of approximately 3,400 students (3,100 full-time equivalents (FTEs)) studying 43 separately identified fields on programmes leading to certificates, diplomas, degrees and honours degrees.

The Oxford Polytechnic Modular Course has grown up with the 'container revolution' in higher education. Initially a source of suspicion, or just plain confusion to those responsible for its approval and validation, as well as to a large section of its potential clientele, it has now become something of an ideal type, closely studied by other institutions concerned to move their own course provision in the same direction. This volume of essays on how the Course has developed and how it currently operates is intended both to assist in the process of sharing novel and good practice and to warn against the over-eager acceptance of Oxford policies and practices as a model. Its authors are keenly aware, and have striven to demonstrate, just how much decision-making in educational development arises from institutional and contextual features which are hard, if not impossible, to replicate. They hope, none the less, that the accumulated experience of 15 years of life of this particular Course, still often referred to as 'experimental', will prove of value to teachers and administrators elsewhere.

'Unit credit' or 'credit accumulation' schemes in higher education are not, of course, new or unusual viewed internationally. The vast majority of North American undergraduate course schemes are of this type, while closely integrated, linear courses with little student choice such as the St John's College programme or the Chicago 'Great Books' scheme are themselves the controversial 'experiments'. In Great Britain the ambitious plan of the Open University relied crucially on offering an array of course units for students to gather selectively for their degrees while, in a more traditional institutional setting, London University science degrees have for many years relied on students accumulating a profile of successfully completed discrete units.

In its simplest sense, then, 'modularity' implies no more than the division of a course into separate elements, each presented to the student as such, normally with separable aims and objectives and a self-contained assessment scheme. In educational code, however, it frequently means more, and doubt is often cast on those schemes which have used the term merely to reflect a crude division of a more conventional scheme (often an existing scheme, where the move is characterized as a reform) without a commitment to other principles. At Oxford the following three principles are regarded as fundamental to the educational philosophy of the modular course.

- *Credit accumulation*
  The compilation by the student of a programme of passes in individual modules ('credits') with the goal of a particular award. Frequently, if not for the majority of his/her time, the student will study these modules alongside students with other qualification aims (in terms of both title and level).
- *Progressive assessment*
  Modules are assessed upon completion of the academic work which they require. With the exception of some elements ('synoptic' modules or examinations) which test skills and comprehension over the range of material which might appear in several modules taken by a student, the extent to which the student has met the aims and objectives of a particular module is assessed and recorded immediately upon its completion.
- *Responsibility and choice*
  Students within the scheme have choices at several levels: of subjects to study (including combinations of main subjects not readily available

elsewhere, for example across the arts and science divide); of qualifications at which to aim; within the range of modules offered by a subject or field; and of modules from other subjects which may be used to complete the requirements for awards. Choices are made within the regulations for particular qualification aims and as set by particular fields; there is also a developed system of academic counselling and checking of the student's programme. However, the programme as finally approved is also confirmed and registered by the student, whose personal responsibility for it is taken seriously.

Figure 1.1 shows how these principles are put into action in the Oxford case. The unit upon which awards are based is the single module, usually taken over one eleven-week term (the course also permits some double modules, which are twice as large in terms of student hours expected and carry two credits). Students in Stage I, which must be completed in two years (it is normally taken in one year by full-time students), are recommended to take 12 such modules and students in Stage II, which must be completed in five years for BA and BSc students and seven years for BEd students (the normal full-time programme is over two and three years respectively) are recommended to take 10 modules in each year of full-time study. The coherence and progression of a student's study is ensured by the requirement to pass modules which are *compulsory* in Stage I (normally 3 for a single field and 6 for a double field) and either *compulsory* or *acceptable* in Stage II (for example, for the BA or BSc honours degree 16 of the 18 modules must be so designated, including at least 7 for each single field).

The classification of a student's honours degree is based on the average mark of the best 18 credits (best 25 in the case of the BEd honours) taken in Stage II. These can include up to two basic modules (normally taken by field students in Stage I) from fields other than the student's own and from which an exemption could not normally be claimed on the basis of some element of prior study. A further regulation prevents the award of an honours degree by attrition: a student extending his or her course beyond the normal period of full-time study has a limit on the number of modules which may be attempted while retaining honours eligibility (21 in the case of the BA or BSc and 32 in the case of the BEd). Further details are available in the *Modular Course Regulations* (Appendix II, p. 139).

The basic structure of the Course reflected in the *Modular Course Regulations* has been unchanged since the original design in 1972. The only significant changes were made to it in the course of resubmission and revalidation in 1984–85, when the regulations were adjusted as follows: a slightly higher threshold for passing Stage I; a higher proportion of the student's normal Stage II programme counting towards classification; a redefinition of the advanced module (from one-twelfth to one-tenth of the normal effort of a full-time honours student) and greater flexibility in the regulations governing combinations of single fields.

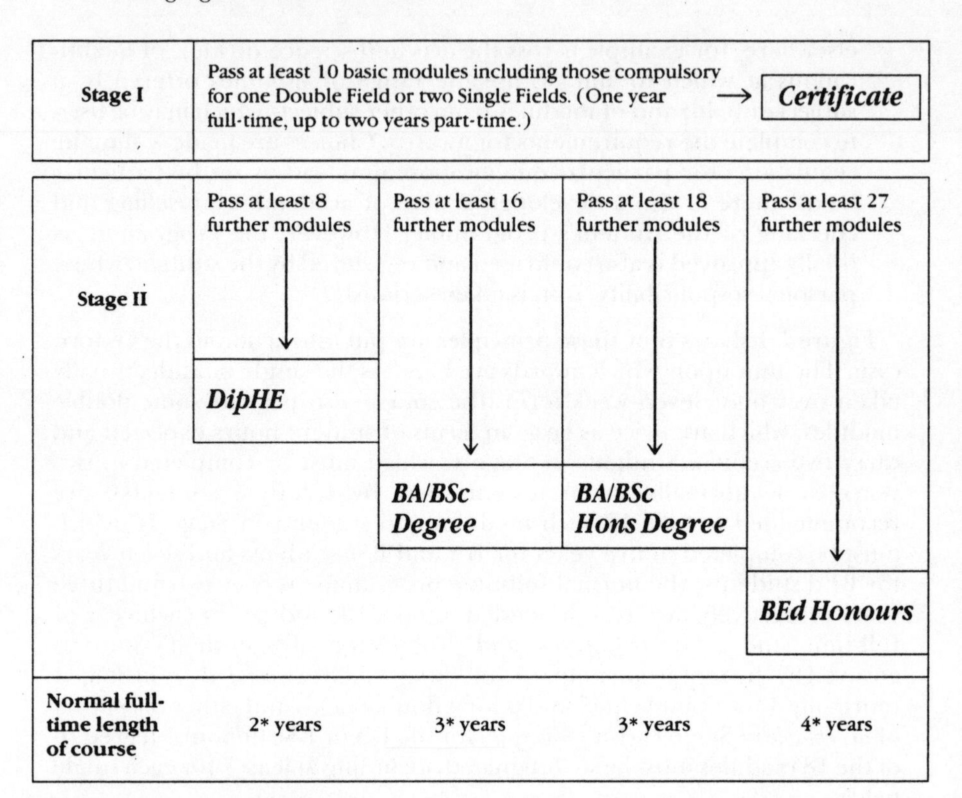

* Add one year for programmes including the single language fields or for the Catering Double field.

BEd for Qualified Teachers – Pass 1 compulsory basic module, and at least 11 for a degree or at least 16 for Honours.

Diploma in Advanced Study – Pass at least 7 modules, at least 6 of them advanced. This is a course for students already possessing some higher education qualifications who wish to take the equivalent of one year's full-time study for personal or vocational reasons.

Figure 1.1 Course structure and types of award

It will be apparent that this structure depends upon and aims to achieve a considerable measure of student control over the educational experience. Unlike the conventional model of the degree course, where a tightly defined progression of study is followed by a straightforward cohort of students who are assessed predominantly at the end of the whole course, the obligation to design and integrate the scheme moves from the staff to the student. *The* Modular Course is, in this sense at least, a misnomer. The modular scheme is fundamentally an organizational device which, while requiring appropriate rigour for the achievement of recognized academic awards, allows for an immense variety of individual pathways to their achievement. There are potentially as many 'Modular Courses' as there are students registered on the scheme.

Advocates and critics of modular course design have contributed to what is by now a fairly sophisticated balance sheet of advantages and disadvantages. The following elements frequently recur:

Flexibility

For	Against
In addition to the elements of choice of courses and units outlined above, the scheme enhances opportunities for transfer and exchange, allows for a varied pace of study, and most importantly enables students to make constant adjustments to their programmes as interests and abilities develop. At Oxford one-third of students make some change to their programmes every term.	In addition to educational objections to a wide range of choice, on which grounds modular courses are often accused of being 'pick-and-mix' or 'cafeteria-style' courses, lacking coherence and progression when measured against conventional schemes, critics can point to circumstances where flexibility can be more apparent than real. Logistical constraints, such as timetable congestion or split-site operation, can often thwart educational objectives.

Economy

For	Against
By offering subject-based units to students on a wide variety of potential programmes, modular courses can achieve significant economies, especially through common teaching and the resulting economy of group size. Small group work, particularly with advanced students, may thus be protected even in circumstances of increasing pressure on staff–student ratios.	Economies may be harder to achieve in practice than in theory; in particular most institutions have inherited patterns of accommodation which make large group teaching on a wide scale difficult to achieve. Similarly, economies on the teaching side may be more than offset by the complex demands of administration and the potentially heavy demands of assessment and examination.

Progressive assessment

For

The student experience of progressive assessment influences programme choice in a healthy and constructive way. Knowledge of 'how I am doing' assists students in working to their strengths, compensating for weaknesses and, most significantly, potentially revising their qualification aims on an informed and realistic basis.

Against

Progressive assessment can lead to 'tactical' as opposed to principled choices and, as a consequence of averaging individual module marks and grades, can lead to insufficient attention to a student's final level of achievement (the 'exit velocity' referred to in many final examination boards).

Careful design

For

In a large multi-subject environment it is important for the left hand to know what the right hand is doing. Module descriptions are written to an agreed formula (specifying, in the Oxford case, relationships with other modules, including level and prerequisites; general educational aims; teaching and learning methods, including a breakdown of likely student hours; course content; and the assessment scheme). They then become public documents available to staff and students and susceptible to regular checks for their academic respectability, potential overlap with other modules and potential use for new programmes or fields. Significant changes are similarly subject to public validation and recording.

Against

Keeping course information at this level generally available is time-consuming and expensive. Public currency of this kind can also lead to conservatism in course design or 'playing safe'. The dangers of over-prescriptiveness and/or superficiality are also difficult to avoid.

The student record and transcript

For

Students completing a qualification on the Modular Course generally have more authenticated information about their educational experience than other graduates or diplomates, both in terms of subjects studied and levels achieved. At Oxford it is strongly hypothesized that the experience of negotiating a programme of study in this amount of detail, and of being able to demonstrate the outcomes to employers, explains in part the success of Modular Course graduates in the job market.

Against

Alternatively the transcript can be read as confirming the patchiness and lack of coherence of the student's course ('islands of knowledge in a sea of ignorance' is one extreme characterization of a list of modules passed by a student). By itself the student's termly record or final transcript cannot confirm that he or she has achieved the intellectual maturity or special skills associated with graduate status in various fields.

Public understanding

For

Modular courses tune in well with current concerns that higher education should be more 'flexible', 'interchangeable', and recognize the achievements of students in non-traditional educational environments; hence the boom in schemes for credit transfer and exchange, for assessment of prior and experiential learning, for distance learning, and for the recognition of non-standard entry qualifications. The 'negotiable' framework of the Modular Course includes opportunities for exemptions and admission with advanced standing as well as for remedial and 'balancing' study. It also allows for 'lifelong' education, especially through the use of the ladder of intermediate awards.

Against

Public enthusiasm and endorsement, especially at the rhetorical level, should not be allowed to disguise the extent to which such courses continue to confuse groups such as employers, schools, colleges and careers advisers. For example, the progress made since the early 1970s in establishing their credibility with validating bodies, particularly the Council for National Academic Awards (CNAA), meant little when public sector institutions were confronted with the national planning framework of the National Advisory Body (NAB) from 1983 onwards. Currently large public sector institutions which offer such courses are girding themselves for another painful round of explanation and potential conflict with the Polytechnics and Colleges Funding Council (PCFC) brought into being by the Education Act of 1988.

The records of two students on the Oxford Course set out below show some of the claimed advantages being realized.

> *Student A*
> She is a mature student part way through her course (she eventually achieved an upper second honours degree). She entered the course, with standard matriculation qualifications, after her children had begun school. In the second term of 1983–84 she became pregnant again and shifted to a part-time mode of study. She then took three terms approved leave before returning to study part-time in September 1984. By the summer term of 1985–86 she was ready to return for a final term of full-time study to complete her degree requirements (see Figure 1.2).
>
> *Student B*
> This is a straightforward case of industrial secondment. She entered the course part-time in September 1980 on the basis of a good HND in computing. During the next ten terms her employers supported her for eight part-time and two full-time terms, matched with their requirements, before she graduated with first class honours in July 1984 (see Figure 1.3).

This Course was broadly in advance of many of the aspects of the container revolution in higher education implied by the general acceptance today of modular course design as appropriate and in some cases necessary for the achievement of particular goals. As the contributors below indicate, the Oxford Modular Course is no longer unambiguously a brand leader and in certain respects (such as the development of distance learning and independent study) may now be seen to lag behind. None the less it does represent a formidable example of an institutional commitment to educational innovation and a valuable source of experience against which other plans and other achievements may be tested.

Name	Home address	Title	Forenames	Local address
Number				
Date of birth 08/06/52				
Date of entry 20/09/82				
Mode of study Full-time				
Fields Sociology/Law				
Senior Tutor/s (ST) or Field Chairmen (FC) Dr F		G432	(FC)/Mr G	
Personal Tutor Dr F		SOC ST G432	H499	Entry qualifications

Stage I						Stage II											
82	1		57011	INT VIS STU	C	83	1		61121	PERSP EDUCN	C	84	2	LW	79222	TORT	B+
82	1		57061	INT GRAPHICS	P	83	1		79041	LEGAL PROCES	B						
82	1		61011	CHILD DVLP	A							84	3	SO	78715D	INDUST SOCS	A
82	1		64041	EARLY LRN NF	B+	83	2	SO	78214T	SOC THEORY	B+	84	3	LW	79713	LEG CONTROVS	B+
82	2		57052	INT PHOTOG	B	83	3		79055D	INTRO LAW	B+	85	1	SO	78131	SOCIAL POL	B+
82	2		61022	THEOR BAS ED	B+							85	1	LW	79514D	LABOUR LAW	
82	2		64052	A & D 5–13	C	83		Terms 2 & 3 Approved time out – course extended				85	2	LW	79514D	LABOUR LAW	
82	3		57077	DRAWING STD	C												
82	3		61033	DIS ED + CUR	B+	84		Term 1 Approved				85	3	SO	78909	SO INTER DIS	
82	3	SO	78013	INTRO SOCIOL	B+			time out – course extended				85	3		79123	CRIME/SOCIET	
82	3	SO	78025D	INDUST & SOC	B+							85	3	LW	79233	CIVIL OBLIGS	
						84	1	LW	79211	CONTRACT	B	85	3	LW	79909	LW INTER DIS	
						84	2	LW	79114D	CRIMINAL LAW	B						

Qualifications expected:
BA honours
Qualifications awarded:
25 07 85 DipHE; not conferred

Summary
Stage I: Passed 12 with an average of 57%.
Stage II: Passed 15 with average of 63% (65% over best 13); adv 12, basic 3; acceptable: SO 6, LW5, either 0, total acc 11. Total taken 15. Transition regulations apply to you (details in Leaflet M13).
You must pass at least 20 credits for Honours including at least 17 acceptable for your two fields

Figure 1.2 Student A – combining part-time and full-time study as a mature student

Name — Home address — Title — Forenames — Local address

Number

Date of birth 03/02/56

Date of entry 26/09/79

Mode of study Part-time

Fields Mathematical Studies/Computer Studies

Senior Tutor/s (ST) or Field Chairmen (FC) Mr D G — TER13 — (FC)/Mr A G

Tutor Mr P J — MS + CM TER4 — H489 — Entry qualifications

Stage I						Stage II											
79	1		84001	BSC MATH MD	EX	80	1	MA	84131	CALCULUS	A	82	1	MA	86131	MATHS STATS	A
79	1	MA	86031	STATS CONCEP	EX							82	1	CO	87111	CATA STRUCTR	A
						80	2	MA	84112	COMP METH	A						
79	2	MA	86022	MATH METHOD	EX							82	2	MA	86152	STATS INFER	A
79	2	CO	87002	INTRO COMP	EX	80	3	MA	84123	LIN ALGEBRA	A	82	2	CO	87142	COMP ARCH	A
79	2	CO	87062	CATA PROCESS	EX	80	3		87025	COMP PROG	A						
												82	3	MA	86183	STAT MODELS	A
79	3		84013	INTRO MATHS	EX	81	1	CO	87151	SYSTEM PROG	A	82	3	CO	87213	COM/MULT SYS	A
79	3	MA	86013	FOUND MATHS	EX							82	3	CO	87253	DATA BSE SYS	A
79	3	CO	87043	COMP-O-A-L	EX	81	2	CO	871240	PROGRAMMING	A						
						81	2	CO	87132	ANAL/DES	A	83	1	CO	84241	DATA COL/AN	A
						81	3	MA	86163	CALCS VAR	A						
						81	3	MA	86193	CRD DIF EQU	A	83	2	CO	87999D	CO PROJECT	A

Qualifications expected:
BSc honours degree
Qualifications awarded:
05 04 84 BSc honours class 1

Summary
Stage II: Passed 21; average over 21: 85%. Progressive average over best 18: 87%. Total taken in Stage II: 21. Advanced 20, basic 1. Acceptable: MA 9, CO 10, either 0, total acceptable 19.

Figure 1.3 Student B – an example of industrial secondment

2
Academic Validation and Change

David Watson

Principles

This chapter considers the history and development of the Oxford Polytechnic Modular Course between 1972 and 1988 from the points of view of academic innovation; the process of validation and professional recognition; the 'fit' with institutional and national planning; and the means of effecting change. Among the points considered are the following:

- *Academic innovation*
  The case for and circumstances of creating the Modular Course. The principal means of growth and development including accretion, redefinition and invention.
- *Validation*
  The changing circumstances of validation by the CNAA, including before and after the Interfaculty Studies (IFS) Board and the impact of delegated authority and accreditation.
  The achievement of professional recognition of fields and combinations of fields.
- *Academic planning*
  Contributions to a Polytechnic development plan. Academic impact of the NAB planning process.
  The prospects of PCFC.
- *The process of change*
  Reaction to various types of initiative: internal and external; 'top-down' and 'bottom-up'.
  Relationship of field-specific (local) and course-wide (central) changes.

The principles underlying course development were established in a set of aims drawn up in 1972 and reaffirmed by the Modular Course Committee (MCC) in 1978 and 1983. The original statement was as follows:

> Experience over the last ten years (1964–73) has convinced the staff of this Polytechnic of the need for a more flexible type of course than is

> commonly offered to students. In particular, it is felt that the student needs to be allowed some freedom to choose the elements of his own course within the disciplines concerned and to be given the opportunity to spend a proportion of his time working outside his chosen disciplines, for which appropriate credit should be given in the assessment of his degree.
>
> To achieve this flexibility, the modular structure has been adopted in the belief that it provides the maximum amount of choice for the student compatible with the maintenance of academic standards. The student is given the opportunity either of broadening his areas of study so that he learns some of the principles and methods of several disciplines or of concentrating on the narrower range in order to become a specialist after suitable experience, training or further study.

The ensuing 15 years have been characterized mainly by a series of opportunities for or constraints upon developing theses principles. Among the former have been the national trends towards acceptance of credit accumulation and transfer and all that they imply; among the latter have been the effect of a tightening national steer on undergraduate student numbers in particular subjects. In 1983, however, the MCC was able to make the following confident series of resolutions to the CNAA.

(i) to re-affirm the principle, inherent in the content and design of the Modular Course, of offering to students a broad-based, flexible course of study in which they participate actively through *choice* of fields, within fields and of study outside their main field. They may, for example (to paraphrase the original submission) either broaden their areas of study in order to learn the principles and methods of several disciplines or concentrate on a more extensive study of a single discipline.

(ii) to exploit more fully the opportunities provided by the unit credit system to offer these studies on *undifferentiated full-time, mixed-mode, and part-time routes*, and hence contribute to local and national needs for initial undergraduate courses and continuing education.

(iii) to build on the established strengths of subject areas and seek to extend *interdisciplinary cooperation* in teaching, learning and research.

Together these principles attempt to establish what is distinctive about the Modular Course. Other, more conventional courses have been faced with similar challenges and restrictions over the past decade. One important theme in the remainder of this chapter is the extent to which a modular course design has made it easier or harder to seize opportunities and work within constraints. First, however, it is necessary to establish the overall record.

The record

The following summary of the chronology of approval of and recruitment to Fields illustrates the steady development of the Oxford Polytechnic Modular Course since September 1973:

1973

Applications	799
Admissions	75
FTE students	75
Fields	7

Course (approved by CNAA October 1972) starts with double fields in environmental biology, geology & environment, and human biology, and single fields in biology, mathematics & computer studies, geology, and physical sciences. (The Polytechnic Diploma in Book Publishing is also integrated with the scheme.)

1974

Applications	1,296
Admissions	170
FTE students	243
Fields	16

Course extended to include arts and humanities single fields in English, French literature, geography, German literature, history, history of art; also psychology, anthropology; double field in physical sciences and combinations of arts and science fields.

1975

Applications	1,874
Admissions	281
FTE students	481
Fields	20

Course extended to allow for BEd degrees and honours degrees and Diploma in Higher Education (the latter retrospectively from 1974); education approved for all degree and diploma programmes; also art & design for BEd only and music for BEd and DipHE only; single fields in catering studies, and food & nutrition (approved for BA and BSc only).

1976

Applications	2,890
Admissions	353
FTE students	717
Fields	25

Introduction of single fields in French language & contemporary studies, and German language & contemporary studies; DipHE programme for food & nutrition, French, and German language & Contemporary Studies.

1977

Applications	2,909
Admissions	371
FTE students	870
Fields	25

No major course developments.

1978

Applications	2,869
Admissions	462
FTE students	1,088
Fields	29

Introduction of single fields in economics, law, sociology, and politics; part-time and mixed-mode regulations come into force (approved June 1978).

1979

Applications	3,557
Admissions	618
FTE students	1,361
Fields	32

Scheme reapproved by CNAA for five years including new single fields in environmental biology, honours status for food & nutrition, catering and new double field in catering; new single field in visual studies replaces art & design for entire scheme; introduction of BEd and BEd (honours) for qualified teachers.

1980

Applications	3,888
Admissions	582
FTE students	1,569
Fields	33

Previously prohibited combinations of the two French fields and the two German fields approved; also two new single fields in computer studies, and mathematical studies replacing mathematical & computer studies.

1981

Applications	4,565
Admissions	589
FTE students	1,716
Fields	35

Named BSc degrees and honours degrees in environmental biology, human biology (awarded retrospectively from July 1981), earth sciences (replacing double field in geology & environment), and physical sciences introduced; also new fields in accounting & finance, and cartography;

food & nutrition field retitled as food science & nutrition. Polytechnic granted delegated authority to make certain types of change to the course without prior approval of the CNAA.

1982

Applications	7,514
Admissions	753
FTE students	1,980
Fields	35

New single field in musical studies replacing music and available for all degree and diploma programmes introduced.

1983

Applications	10,000
Admissions	691
FTE students	1,997
Fields	36

New single field in publishing replacing Polytechnic Diploma in Book Publishing and available for all degree and diploma programmes introduced. Scheme reapproved by CNAA indefinitely subject to progress reviews. General Polytechnic Diploma in Advanced Studies introduced.

1984

Applications	11,000
Admissions	711
FTE students	2,180
Fields	36

Revised regulations for Stage I introduced. Also 'independent study' modules and new examination arrangements as agreed with CNAA.

1985

Applications	11,000
Admissions	759
FTE students	2,333
Fields	36

Revised regulations for Stage II introduced, and 'transitional regulations' for students entering Stage II of the course before September 1985. New award of certificate established. Major review of computer studies and submission of reports on physical sciences mean that all fields are approved indefinitely by CNAA. New timetabling scheme introduced.

1986

Applications	15,120
Admissions	799
FTE students	2,381
Fields	36

New fields in microelectronic systems (single), tourism (single), cell biology (double), languages for business (double) for September 1987; also planning studies (single and double fields, and double field with RTPI exemption) for September 1988. Overall report to CNAA on information technology across the Course.

1987

Applications	15,575
Admissions	938
FTE students	2,658
Fields	39

New single fields in business administration & management, physics, and chemistry approved for September 1988 (the latter two replacing the double field in physical sciences); five double fields comprising the integrated degree in nursing and midwifery approved for start in September 1989.

1988

Applications	20,000
Admissions	923
FTE students	2,857
Fields	43

Plotting the progress of students by cohort shows the effect of new fields upon the Course, and also of the possibility of mixed mode study (Table 2.1). The analysis also confirms the strong record of the course in retaining students and seeing them through to their final qualifications. Table 2.2 links the number of students who could have achieved a degree award by the summer of 1987 with the number of awards actually made. It excludes intermediate awards (the Certificate and Diploma) and those students who have extended the course for reasons other than failure (Table 2.2).

It should not be forgotten that the Modular Course has expanded in an expanding Polytechnic. For the quinquennium 1982–83 to 1986–87, i.e. after the major phases of absorption of other large courses, it has remained just below half of the FTE students in the Polytechnic. From September 1988, particularly through the inclusion of planning studies, it will move to just above half (Figure 2.1).

Student registrations on the course by mode show a small increase in both sandwich and part-time modes which, interestingly, has been matched by a corresponding decrease in such registrations elsewhere in the Polytechnic so that institutional figures are roughly stable over the period from 1982–87. The phenomenon reflects not only the difficulties of part-time recruitment in an area like Oxfordshire which lacks a large urban and industrial base, but also the fact that the Modular Course has a better chance of retaining and increasing such students than conventional courses (Figure 2.2).

Table 2.1 Progress by cohort 1973–87

	Year of entry														
	1973	*1974*	*1975*	*1976*	*1977*	*1978*	*1979*	*1980*	*1981*	*1982*	*1983*	*1984*	*1985*	*1986*	*1987*
Year of Study															
1973	75														
1974	71	170													
1975	65	147	272												
1976		129	240	353											
1977		20	214	265	371										
1978			28	293	311	462									
1979				19	305	419	618								
1980				6	80	367	535	582							
1981					11	89	509	518	589						
1982					2	15	161	508	541	753					
1983						4	15	165	493	662	691				
1984							2	20	152	574	608	711			
1985								8	19	224	587	613	759		
1986								1	1	16	177	626	673	804	
1987										2	12	207	650	715	958
Awards in Year	1975–76	1976–77	1977–78	1978–79	1979–80	1980–81	1981–82	1982–83	1983–84	1984–85	1985–86	1986–87			
	57	107	207	261	271	376	444	481	532	623	672	586			

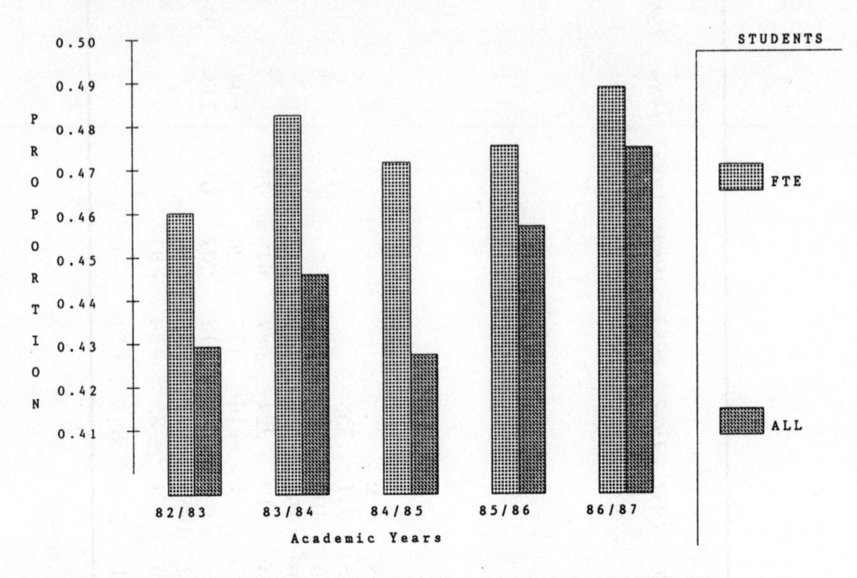

Figure 2.1 Modular Course as a proportion of all courses 1982–87

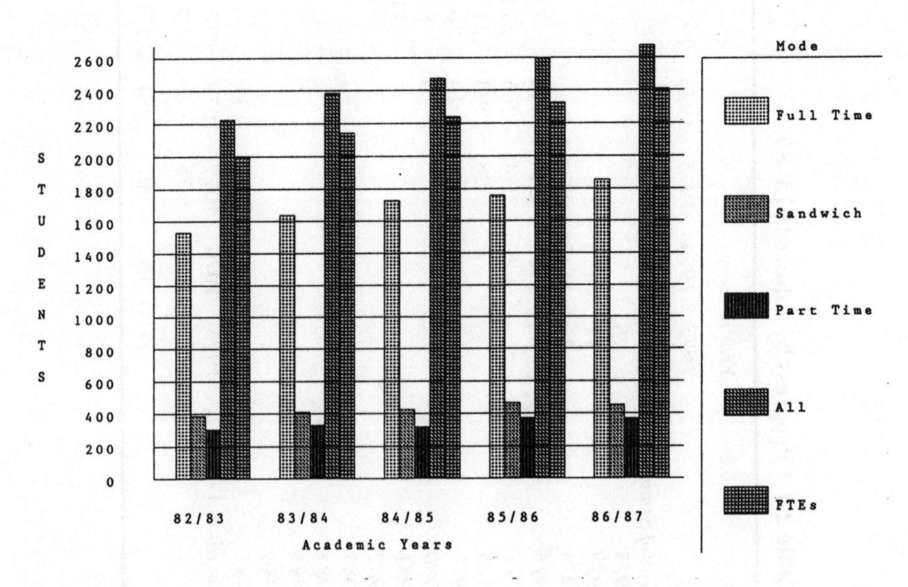

Figure 2.2 Modular Course registration by mode of study 1982–87

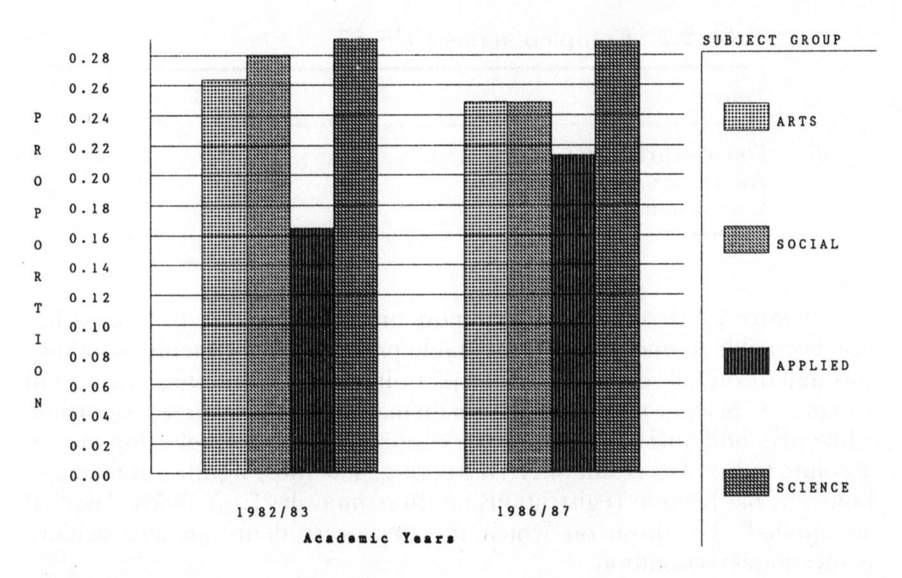

Figure 2.3 Module registrations by student group 1982–87

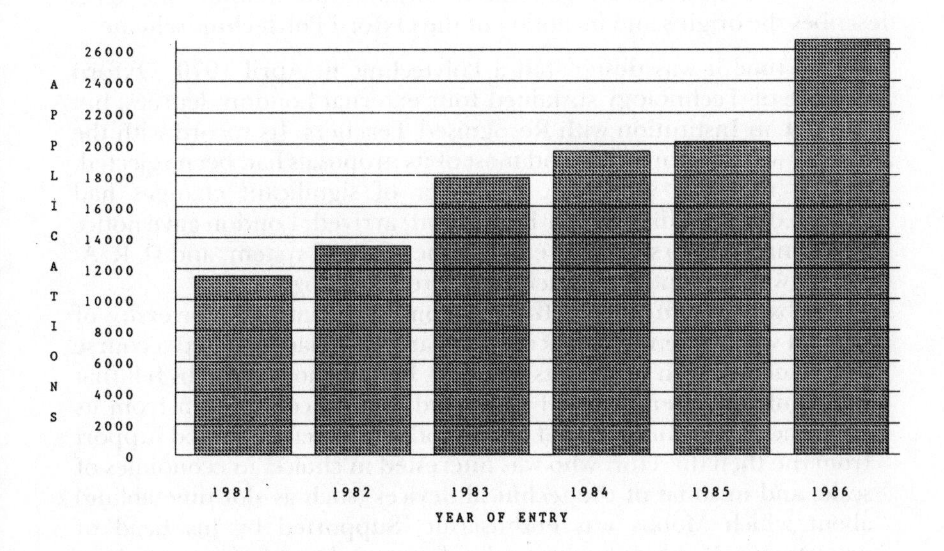

Figure 2.4 Oxford Polytechnic: applications for full-time courses 1981–86

Table 2.2 Completion rate 1976–87

Years	*1976–87*
Potential graduates	5,522
Awards (excluding diplomas)	4,617
Completion rate %	84%

As Figure 2.3 (from the same quinquennium) indicates, the Course has also been able to maintain a fairly stable pattern of enrolments by subject through the initial years of NAB-led planning (an issue discussed further in Chapter 3). Science registrations remain marginally the largest proportion, while arts and social sciences have dipped proportionately (but not in absolute terms) as a result of NAB steering. The most significant increase, however, has been in registrations on those modules from fields classified as 'applied', i.e. those on which the degree student can also achieve professional recognition.

Finally, the Course remains an extremely popular option for applicants to an extremely popular Polytechnic. The data on applications at the head of this section can be checked against the information in Figure 2.4 for the Polytechnic as a whole.

Academic innovation

This is how Ernest Theodossin, the historian of the modular movement, describes the origins and inception of the Oxford Polytechnic scheme.

> At the time it was designated a Polytechnic in April 1970, Oxford College of Technology sustained four external London degrees, but was not an Institution with Recognised Teachers. Its record with the CNAA was unimpressive, and most of its proposals had been rejected. Before the year was over a number of significant changes had occurred: a new director (Dr B. B. Lloyd) arrived; London gave notice that it intended to stop bulk entry to the external system; and D. R. A. Mobbs was appointed a principal lecturer in biology.
>
> Mobbs had returned to Britain from a post at the University of Zambia where the vice-chancellor, a Canadian, had imposed a course unit structure with four units per year, but no choice. Mobbs felt that the Zambian system worked badly and that he could learn from its deficiencies. His commitment to the modular structure gained support from the then director, who was interested in choice, in economies of scale, and in some of the technical devices (such as slot timetabling) about which Mobbs was enthusiastic. Supported by his head of department, Mobbs was given relief from teaching duties to work on course development.

> With the impending withdrawal of London validation, the institution was receptive to new ideas. Mobbs himself felt that he was 'the new broom' and that 'things were working' for him. He was seeking to develop the widest possible range of courses, so that each person 'could reach his potential in academic, professional and vocational work' (which he saw as 'not separate') in such a way that changes of direction would be possible at many points. However, had he put forward a radical package, with organisational, administrative and hierarchical implications, he might have encountered resistance. Instead, the sketch was clear only in his and the director's minds, and what was done was to begin in the limited area, biology and geology. Later, Mobbs was joined by other subject groups.
>
> Mobbs started work on the project in January 1971; in May 1972 the proposal was sent to the CNAA; it was approved after a one-day visit in November 1972 and the first intake arrived in the autumn of 1973. The Oxford Modular Course was the first multi-disciplinary modular course validated by the council.

This outsider's account rings true for insiders in two ways. First, it emphasizes the extent to which the Oxford Polytechnic Modular Course was created *de novo*, with all that that entailed for the enthusiasm of colleagues and the mystification of outsiders. Second, it accurately acknowledges the source of the scheme. It remains true that all of the major parameters of Course regulations and management remain today as they were established by David Mobbs; all that has changed is the values given to certain parts of the equation. Theodossin also implies a third element here, that the scheme began and developed with a wide range of institutional consent precisely because of features of its initial design (he points, for example, to the choice of a term-long unit and the control retained by staff over field changes, if not module changes, by students). His conclusion is that 'the Oxford scheme allows the maximum manoeuvrability consistent with not arousing the anxieties of vested interests among the educational establishment'.

This is not how it has always felt. Some of the early groups joining the Modular Course as they faced the same problem of replacing London external degrees (humanities and social science) felt the strong hand of management as they exercised their 'academic' decision to join. (There is a story, which it has proved impossible to verify from the record, that Lloyd opened one meeting with these staff by declaring 'this is not a democracy'.) Other groups, especially more recently, were enthusiastic rather than reluctant. Those responsible for initial teacher training, for example, were apparently relieved that the Polytechnic, upon its merger with the Lady Spencer Churchill College of Education in 1975, was not intending to leave them as a monotechnic island in the interdisciplinary sea of the Polytechnic, particularly when it remained possible (as confirmed by subsequent experience elsewhere) that isolation would mean a low ranking in whatever prioritization exercises the Polytechnic might be forced to undertake.

Essentially there have been three methods by which new subject areas have joined the Modular Course. Numerically easily the most significant has been the process of *accretion*, whereby existing courses have joined the scheme, accepted its rules and conventions and added their subject contribution, in the form of fields, to what is already on offer. This is the process experienced by arts, languages, social sciences and most recently planning. The act of joining the scheme in this way is of major significance to the delivery of courses recast in this way. Until the 1983–84 submission to the CNAA (discussed below), for example, several new areas felt constrained and unconvinced by aspects of the operation (especially timetabling) which made perfect sense for a full-time course in science but considerably less for other kinds of demands. For example, the history seminar had to struggle to find a two-hour block and the language class could never achieve the desired continuity by organizing its contact time over three, let alone four days of the week.

Second in importance until recently was the process of *redefinition*. Here the Polytechnic took established courses, often at a different level, which upon review no longer seemed appropriate or sustainable as part of the institutional profile, and redesigned them as fields on the Modular Course. Examples of this kind include the Polytechnic's Diplomas in Cartography and Publishing (now single fields), and, most recently, the Polytechnic Diploma in Language Studies (now a double field, leading, over three years rather than the four required for honours degrees in other language fields, to a BA degree only). Motives here have varied, and included a changing perception of the needs of employers, course costs which were hard to justify in comparison to other areas of the Polytechnic and the need for degree designation in order to attract mandatory student grants.

Third, there are the instances in which the array of subjects and modules across the Course, combined with external pressures and challenges, have facilitated the *invention* of new fields, and their development at significantly less cost than would have been the case without the modular scheme. Examples here are tourism, business administration & management, and microelectronic systems. Each draws significantly upon previously existing courses and each attempts to fill a gap in the current national provision of undergraduate education.

Finally, there is, of course, the possibility of hybrids between two or more of these methods of course development. The five double fields (each related to a professional register) which comprise the degree in nursing and midwifery involve redefinition of the existing training scheme, accretion of new subject areas in health-related studies and the invention of a new and unique opportunity for intending students.

Validation

The Course as a whole has been presented for validation and revalidation to the CNAA three times. In 1972–73 it was initially approved after some preliminary negotiation and a one-day visit, as set out above. The ease of this initial approval of fields in science, where the subject culture was traditionally more sympathetic to modular course design (including progressive assessment, behavioural objectives and all that goes with them), contrasted strongly with the validation experiences of the first major accretions to the Course in humanities, social sciences and teacher education. Here the subject culture reacted antagonistically, with the result that the initial approvals were hedged around with conditions.

Partly as a consequence the renewal of approval of the whole Course, conducted over two days in 1979–80, was based on a submission in which 'safety first' had been the internal maxim. Strong instructions went out to fields reviewing their progress and preparing proposals that there should be as little change as possible from that which had been originally approved. The event itself, one of the first to be conducted by the CNAA's new IFS Board, was an archetype of the theatre of validation, with 120 people at one stage sitting in the Polytechnic boardroom discussing whether or not the course regulations met Principle 3 of the CNAA's own regulations (about the balance and nature of the course of study).

Left over from that encounter and resolved later in the session was the eventual agreement by the CNAA to the Course being offered on a mixed mode and part-time basis. Here again a feature insisted upon by Mobbs, that the regulations as a whole be amended to become agnostic as to mode, proved timely but initially controversial. CNAA members wished a set of special arrangements to be made for part-time students almost amounting to a separate course. By now virtually all new courses elsewhere follow the Oxford pattern.

Changes were also taking place in the central management of the Course. Mobbs had resigned from his elected post as Dean. A successor (all Heads of Department were eligible for election) had proved hard to find and for a brief period the chairmanship of the Course had been taken over by the Deputy Director, Brian Tonge. Finally, and in time for the 1979–80 visit, a Dean was elected: Brian Clark from the Department of Mathematics, Statistics and Computing. Clark had, however, made it clear that he did not wish to continue for a further term and, as a consequence of an Academic Board working party's recommendation, a permanent post was created and filled externally by the appointment of the author.

This put the central management of the Course on a more secure footing for the important preparation for the second resubmission of the Course in 1983–84. In a two-year period of review from late 1981 the Course as a whole and all fields within it reviewed overall principles, the design of the Course and the specific contributions of subject areas. This large-scale exercise was conducted in the knowledge of considerably greater sympathy on the part of CNAA validators, as reflected in the developing working style of the IFS Board. Indeed, one measure of their growing confidence in the scheme had been the agreement, from November 1981, to allow the

Polytechnic delegated authority to make certain types of changes to the Course without the prior agreement of the Council.

The specific areas of discretion were as follows:

- updating the syllabuses/reading lists;
- varying individual module assessment;
- reorganization of modules;
- variation in prerequisites as a result of the three changes above;
- temporary addition of a module;
- modifications to specific field regulations as a result of approved changes in structure or content;
- adding or dropping a module in a particular subject area;
- introduction of new technology/learning methods which might alter student contact time; and
- amendment of general regulations where necessary for consistency.

Such discretion was granted on the understanding (1) that any changes made to the approved Course regulations were notified to the IFS Board for information; (2) that all changes to the Course were monitored and evaluated for the purpose of informing the next submission for renewal of approval, and subsequent Course review visits; and (3) that all changes to assessment requirements were agreed with the appropriate external examiner.

The effects of this delegation were to give confidence to the Polytechnic's internal processes of review and validation and to create a body of case law about their achievement in time for the revalidation of the Course in 1984–85. By that time six termly reports covering over thirty changes (some quite major, including a wholesale review of the catering fields) were on file.

The internal review in preparation for the 1984–85 validation event was a massive exercise which galvanized the whole Polytechnic (one conclusion from the CNAA's report was that the process of critical appraisal could be overdone). In one sense it might also be said to have laboured to produce a mouse. Principles underlying the Course were reaffirmed, as were the framework of the regulations and the system for delivering the Course (semesters in place of terms, for example, were rejected). The new elements put forward were as follows:

- a redefinition of the advanced module as one-tenth of the notional effort of a full-time honours degree candidate in one year (the basic module remained as one-twelfth);
- a relaxation of the regulations on the combinations of single fields in Stage II (the previous regulation in effect required a 50 : 50 split, now 7/16 acceptable modules was the minimum for one field (the Course rejected the idea of calling this a major/minor system);
- raising the threshold for passing Stage I to 10 modules (previously 8 credits with an average of 40 per cent over the 12 taken was an alternative route);

- increasing the proportion of Stage II over which classification is calculated (18/20 for BA/BSc from 18/24);
- the introduction of independent study modules into those fields wishing to include them, under carefully controlled conditions including prior approval of the work scheme and assessment schedule (by 1986–87 24 fields included these);
- further refinement of the standard module description (which now includes details on general education aims, the assessment scheme, as well as an outline of teaching methods, but no longer requires book lists);
- the introduction of a two-tier examination system (described in Chapter 5 below); and
- the use of field titles, and not just the anachronistic formula 'on a Modular Course' on all degree certificates; this was accepted by CNAA for all but the BEd degree – combinations of single fields being prefaced by 'Combined studies in . . .'.

The CNAA's processing of this submission was radically different from its previous practice. After a preliminary strategic session in London, fields were divided into four cognate groups and each was visited separately for a day on the basis of a volume of field proposals and a critical review. A final one-day visit by a 'core group' of members approved the scheme indefinitely and agreed that the future pattern of review would be on the basis of a rolling cycle established by the Polytechnic. Talk within the institution was of having fought 'the war to end all wars'.

In effect the proposal for future review which emerged from the 1983–84 visits anticipated what the Polytechnic achieved through accreditation by the Council from 1 April 1988. The Polytechnic is now responsible, though its Academic Board, for the processes of both review and initial validation under an accreditation agreement. The cycle of periodic reviews of fields is in place and the first initial validations have been led from within the Polytechnic, including those for physics, chemistry, business administration & management, and nursing and midwifery.

Professional recognition

In addition to the 15 years of dialogue with the CNAA recorded above the Modular Course has been required to negotiate with a series of professional bodies about the appropriateness for professional recognition of study on various of its fields. Progress has been patchy but by the time of writing study on the following fields (often requiring further compulsory elements or specific combinations of fields in addition to the general requirements of the course regulations) carry agreed professional exemption or recognition for graduates:

- Accounting & finance
- BEd
- BEd for qualified teachers
- Catering management (single and double)
- Law
- Nursing & midwifery
- Planning studies
- Psychology

The list comprises the formal recognitions. Students have also been successful in making individual applications (for example in computer studies to the British Computer Society) and, indeed, such applications in the early years often formed the necessary groundwork for institutional recognition. Some areas remain, however, aloof from recognizing modular course possibilities. The slowness of engineering to join the scheme, for example, is explained to a considerable extent by fears of the attitudes of the professional societies (it is hoped that, in time, field combinations such as computer studies and microelectronic systems will succeed in breaching this important wall).

Academic planning

The position of the expanding Modular Course within the planning processes of the Polytechnic causes some major problems and helps to solve some others. The complications arise from the way in which it has to maintain an internal planning dynamic which cuts across the traditional demarcation lines of faculties and departments.

At the broadest level, the Course contributes in a major way to the achievement of the Polytechnic's overall aims. These were summarized by the Academic Board in January 1987 as follows:

1. To provide a distinctive higher education service of undisputed quality which differs from much traditional higher education in the following ways:
   (a) its emphasis on relevance to employment, citizenship and personal development (including the development of critical and rational minds);
   (b) the availability of opportunities to study at a broad range of levels, of course lengths and of modes (part-time, full-time or mixed);
   (c) the possibility of combining different areas of study in ways not generally available elsewhere in the UK;
   (d) its strong links with the local community.
2. To extend higher education opportunities for those who are at present under-represented in higher education.
3. To facilitate the continuing personal and professional development of those in work or seeking to re-enter or change work.

4. Through applied research and consultancy activities and courses, to assist business, industry, cultural organisations, public services, the professions, and the community at large.
5. To contribute to better international understanding by establishing links with institutions abroad, fostering exchanges and admitting students from overseas.

While the Course facilitates meeting all of these aims, it has a particular responsibility in respect of 1b and 1c, 2, 3 and 5 and enables the Polytechnic to make a distinctive contribution to the national system of higher education in these areas.

The difficulties tend to come when these broad aims have to be translated into specific objectives for local areas of the Polytechnic, and in particular into detailed forward planning of student numbers. Details of the latter process are given in Chapter 3 below. However, from the point of view of the strategic considerations in this chapter the following points are relevant:

- The extent to which the Course encourages and allows choice puts a significant degree of *uncertainty* into the planning process. This arises not only from potential changes of field (where student preferences are not guaranteed – the contractual baseline is merely that students are enabled to complete the fields upon which they originally registered) but more significantly from the changing pattern of extra-field choice. Here, despite all of the external constraints the Course has managed to maintain the principle that every student should be allowed to take *any* module for which he or she is qualified and which can be timetabled. To allow these freedoms and pursue detailed academic planning requires sophisticated and reliable forecasting.
- Departments and the fields which they contribute to the Modular Course are not free agents, either in terms of conformity to Course regulations or in being able to project student numbers in isolation. Any change to a field will have some knock-on effect on other areas in that it will modify the overall pattern of student choice.
- There are many initiatives which it is more sensible for the Course as a whole to pick up than for individual fields and departments to plough a lone furrow. Examples would be recruitment strategies (for example, the priority given to local mature entries), aspects of teaching methods (for example, the development of distance learning), or response to local or national surveys (for example, of courses likely to meet training needs).

As a consequence of these complications the Modular Course has to operate a planning and response structure, through its own committees and crucially through a termly meeting of Heads of Departments contributing to the Course, that is not only sensitive to overall Polytechnic directives but also ensures a sensible collaborative response. The complex history of the Polytechnic's involvement with the internal planning exercise directed by the National Advisory Body for Public Sector Higher Education (PSHE) from 1984–85 illustrates this point clearly.

NAB's planning process assigns student numbers to institutions by broad subject categories (initially there were 14 of these, subsequently 19). Each student FTE in each of the categories then entitles the institution to a specific unit of funding. Overall institutional allocations might then be fine tuned (there was a process initially called 'mitigation' which protected traditionally high cost institutions from the full force of change) and there are other criteria to meet such as targets for sub-degree, part-time and mature intakes; but this distribution of packets of money by students identified with subject areas remains the basis of the system. Obviously for the Modular Course, with its large number of students on two-subject degrees and the freedom for all students to select modules outside their main subject areas, it posed severe problems.

These problems were compounded by the Polytechnic's principled objections to the NAB's overnight conversion of institutional responses to hypothetical planning scenarios devised by the DES (involving a minimum cut of 10 per cent) into 'bids'. In the event a system of planning and monitoring of Oxford Polytechnic student numbers based on selecting (arbitrarily) the lower of the two DES course numbers for each student on a combination of single fields was agreed, which enabled Oxford to achieve NAB targets accurately in growth areas, without sacrificing the essential contribution of other areas. (Many of the growth areas remained buoyant precisely because students could combine 'favoured' and 'unfavoured' areas of study.)

To have Oxford Polytechnic, the largest such scheme in the sector, apparently circumventing the NAB's preferred method of disaggregating modular courses was, however, an embarrassment and for the next cycle, commencing in 1987–88, the Polytechnic shifted to an alternative based on 'subject weight'. This mirrored the Polytechnic's own method of calculating student FTEs, by reviewing all of the module registrations (regardless of fields) and consolidating them into subject groups. The Polytechnic had argued convincingly in the first round that to have done so then would have simply involved the NAB in steering its 'service' work around. (For example, to have increased numbers in computing, when this was a 'free choice' of increasing popularity, would have had no effect on the number of students graduating from the field of computer studies.) Given NAB's acceptance of the new base of the 1986–87 numbers as achieved under the former system from 1987–88, a 'subject weight' system is now in place.

At the time of writing, the intention of NAB's successor, the PCFC, to replace NAB plans by individual institutional contracts, is clear. This will satisfy 'market forces' to the extent that the essential link with a centrally agreed unit of resource will be cut. It is less clear how PCFC, which will operate under the conditions of the Education Act of 1988 from 1 April 1989, will organize the detailed tendering which will be required. It seems, however, inevitable that the Oxford Polytechnic Modular Course will once again be forced to revise its planning and monitoring methodology and establish new arrangements to preserve its central academic features.

The process of change

These overall Course considerations may give the impression of a system that is inflexible and resistant to change. At times this is undoubtedly true. Devising a replacement timetabling system for the original scheme invented by Mobbs for a full-time course in science took at least ten months of negotiation and debate. Similarly the overall scheme does set constraints against which some local initiatives (a desire to incorporate more fieldwork into the regular term, for example) battle in vain. However, the Course does include smooth and well understood processes for achieving change, connected with the Polytechnic's system of annual review, which have been extensively used.

The main categories of change, and the processes through which they work are set out in Figure 2.5. Essentially this document brings a process which has been in operation since 1980 into line with the accreditation agreement with the Council. Each term approximately twenty proposals are reviewed and evaluated before agreement by the Academic Standards Committee (ASC) on behalf of Academic Board. They can range from modifications to an assessment scheme to a wholesale review of the structure and aims of a field.

Initiatives can also be taken centrally which are then translated into particular applications by fields. The account in Table 2.3 of the build up of

Table 2.3 Independent study: the build up

Year	*Student registrations*	*No. of fields*
1984–85	9	4
1985–86	136	23
1986–87	133	24

Dividing the registrations in 1986–87 into broad subject areas also shows a fairly even spread across the course.

Subject area	*Student registrations*
Arts and Humanities	55
Science and Technology	29
Social Science	26
Education	23

oxford polytechnic
modular course

M126

INTERNAL VALIDATION - PROGRESS DOCUMENT

Field(s)..

Nature of Proposal or Review ....

....

This document monitors the progress of new course proposals and changes as they receive approval at the various steps in the internal validation process. The specific items to be processed in this way are:

(i) a new field proposal
(ii) a major (quinquennial etc) review of a field or fields
(iii) a proposal for the (a) addition of a module
(b) deletion of a module
(c) substantial syllabus revision
(>25% accumulated over time)
(d) change in prerequisite
(iv) a change in the assessment of a module
(v) a change in field regulations

The Field Chair should notify the Dean and Chair of the appropriate Faculty Group by the end of Week 2 in the term in which it is wished that the proposal be considered. **It is the Field Chair's responsibility to steer the proposal through the committees** and forward it to ASC with copies of this document, the appropriate minutes, reports etc. Proposals which are dealt with within the institution (under discretionary powers granted by the CNAA) will normally take effect no earlier than the next academic year. Proposals in categories (iii), (iv) and (v) will normally arise from field or departmental annual reviews.

Steps 3 and 4 below are alternatives, to be selected by the Chairs concerned. The appropriate minutes should be attached to this document together with the current approved Field diagram and Field list, and the item ticked below as approval is given. Steps 5,6,and 7 will only apply for items in categories (i) and (ii) above.

STEPS

	✓	Minute Reference	Date	Revisions Accepted Chair's Signature
1. Appropriate Field Committees				
2. Appropriate Subject Committees or Departmental Boards				
3. Faculty Review Group				
Modular Management & Review Committee				
4. Combined FRG & MMRC				
5. ASC Assessor's Report		Assessor's Name:		
6. 2nd Internal Assessor's Report		Assessor's Name:		
7. External Assessor's report		Assessor's Name:		
8. Academic Standards Committee				

RMT/JB Jan 88

Figure 2.5 Internal validation – progress document

independent study across the Course would be an example of a course-wide policy which is permissive progressively permeating the Course as a whole.

Returning to the record established at the head of this chapter, the Modular Course has obviously been transformed as it has grown. It has responded to internal and external stimuli and absorbed the tensions between local and central needs and 'top-down' and 'bottom-up' initiatives. The external pattern of validation and planning for student numbers has posed significant problems to be overcome but has not significantly impaired a dynamic experiment in course development.

3

Subjects and Courses: Managing the Matrix

David Watson

Principles

Managing a large, multi-subject enterprise like the Modular Course makes two principal demands upon an administrative and consultative system: first that it can cope with a high degree of complexity and second that there should be clear lines of accountability for decisions made and the responsibility for carrying them out. Both demands are further complicated by the embedding of this large enterprise in an institution, and across faculties and departments, which also have responsibility for non-modular work.

The chosen device for surmounting these problems on the Oxford Polytechnic Modular Course is the matrix. Responsibility for deploying resources in support of courses is separated from responsibility for academic management and maintenance of standards. The former is supplied along an axis controlled by faculties and departments, led by Deans and Heads. The latter is the purview of another axis in which elected Field Chairs are the most significant individuals. Where the two axes meet, particularly in the context of policy initiation or adjudication, a small number of academic staff described as the central academic team have particular responsibilities.

Several important issues arise out of this agreed strategy, including the following:

- the articulation of academic and resource planning;
- the balance of academic and administrative control;
- the demands of briefing and training of staff at all levels;
- the difficulties of supplying quick and flexible response to opportunities and perceived difficulties; and
- the balance between local and central needs and aspirations.

In the remainder of this chapter these issues are tackled in the following way. An outline of the Modular Course Matrix is followed by a discussion of roles and responsibilities of key individuals and groups. To illustrate the type of problems encountered and the extent to which they are overcome the process of forward planning of student numbers is chosen as a test case. Finally, a summary of policy issues raised and discussed during one academic year (1987–88) is used to demonstrate the pattern of contending interests and the extent to which they can or cannot be accommodated.

The Modular Course matrix

Figure 3.1 summarizes the main features of the Modular Course Management System. The vertical axis is headed by elected Deans of Faculties and appointed Heads of Department with immediate control over the distribution of resources. Particular fields may be located entirely or predominantly within each department, and a departmental 'home' is identified for each field to ensure accountability for the adequacy of resourcing and planning arrangements. The academic development and day-to-day management of fields, which may be made up of modules contributed from more than one department, are, however, in the hands of Field Committees headed by elected Field Chairs (the horizontal axis). Overall responsibility to the Academic Board and the Director for the health of the Course lies with the appointed Dean, who heads a small central team consisting of the Assistant Dean, Modular Course Co-ordinator and eleven senior tutors. Meanwhile the Course draws upon central services (such as the library, computer centre and Educational Methods Unit) as well as the administrative resources of the Registry and examinations office.

Individuals

The central team

The central team of academic staff responsible for the management of the Course has grown with the size of the Course itself, although not always as fast as the individuals concerned would have liked. Table 3.1 correlates the course FTE and number of fields with the administrative staff complement within the registry primarily associated with the Modular Course and, in the right-hand two columns, the academic staff.

As explained in Chapter 2, from September 1981 a permanent Dean replaced an elected Head of Department in the role of Dean of the Modular Course. From April 1987 (after a year in which he acted as Deputy Director of the Polytechnic) he was redeployed as Assistant Director and Dean of the Modular Course on the assumption that he would have half of his time from September 1987 available for Course administration. At that time the post of Assistant Dean was created to bring the number of central

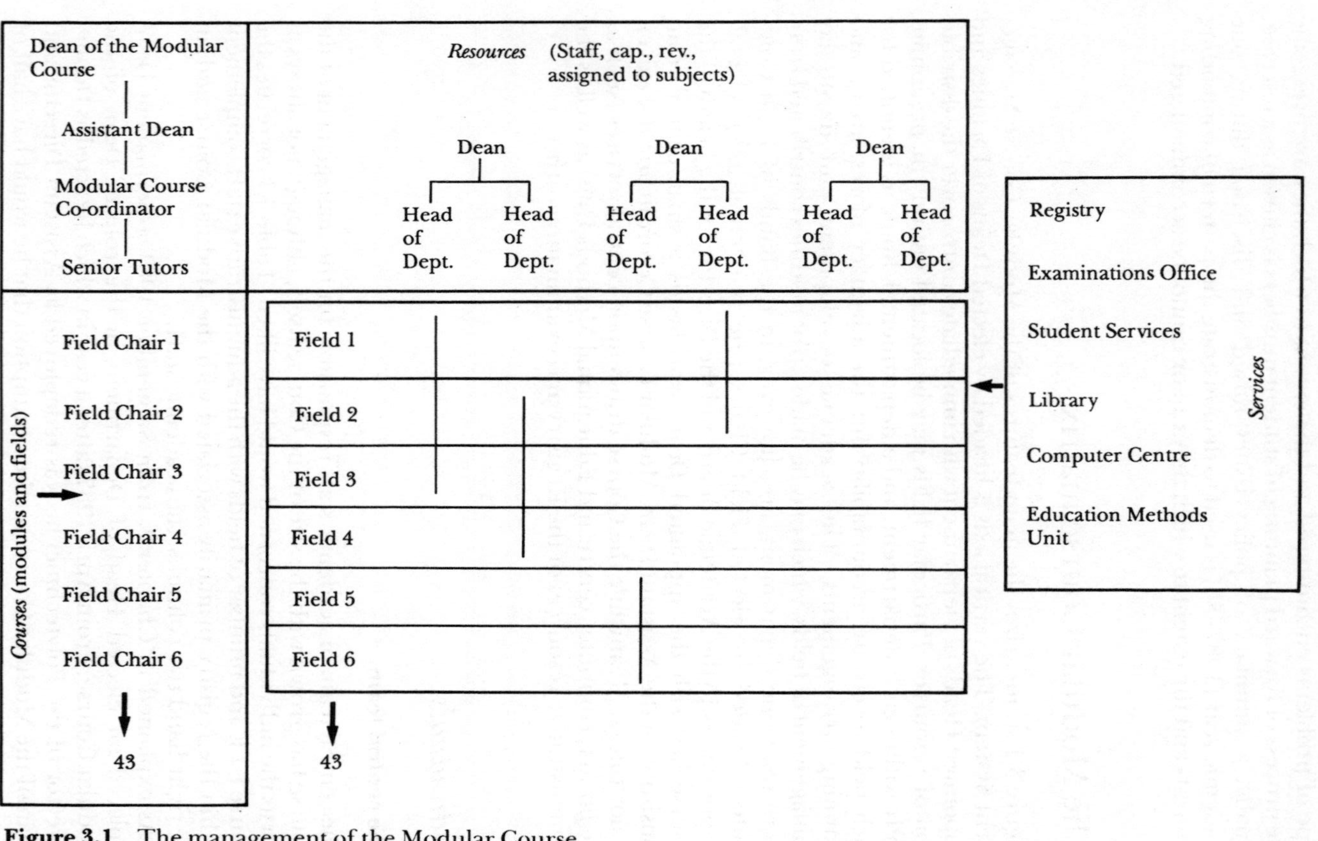

Figure 3.1 The management of the Modular Course

Table 3.1 Central support for the Modular Course 1978–88

Year	*FTE*	*Fields*	*Modular Course administrative staff*	*Modular Course academic posts*	*Senior tutors*
1978–79	1,088	29	6	1	8
1979–80	1,361	32	6	1.5	8
1980–81	1,569	33	6.5	1.5	8/9
1981–82	1,716	35	6.5	1.6[1]	9
1982–83	1,980	35	6.5	2	9
1983–84	1,997	36	6.5	2[2]	9
1984–85	2,180	36	6.5	2	9
1985–86	2,333	36	6.5	2	9
1986–87	2,381	36	6.5	2	9
1987–88	2,658	40	7	2.5	11
1988–89 (predicted)	2,857	43	7	2.5	11

Notes: [1] Permanent Dean replaces rotating deanship
[2] Co-ordinator confirmed as permanent

posts to 2.5 (including the Co-ordinator). The current disposition of responsibilities between the three posts is as follows:

Dean of the Modular Course
As Dean of the Modular Course the Assistant Director/Dean of the Modular Course is responsible for the academic leadership and management of the Course. He has particular responsibility for:

- development of Course provision within the policies of the Polytechnic;
- deployment of the academic management team, including responsibility for staff development;
- liaison with the directorate, Deans of Faculties and Heads of Schools/Departments contributing to the Course; and
- external liaison on behalf of the Course, for example with NAB/PCFC.

In furtherance of these aims he represents the Modular Course on the Academic Board and its subcommittees. He chairs the Modular Course Committee (MCC) the termly meeting of modular Heads of Departments, and the termly meeting of staff and students. He is an *ex officio* member of the Modular Examinations Committee (MEC) and Modular Management and Review Committee (MMRC).

Assistant Dean of the Modular Course (new post from 1 September 1987)
The Assistant Dean is responsible through the Dean for the day-to-day management of the Course. He takes special responsibility for:

- detailed forward planning, including the implications of student target places;

- the review and validation system;
- liaison with validating bodies (for example, CNAA) and external bodies concerned with credit transfer (for example, South East England Consortium for Credit Transfer (SEEC));
- student progress in Stage II;
- preparation of materials for examination committees; and
- student exchanges.

In furtherance of these aims he represents the MMRC on the Academic Standards Committee (ASC). He also chairs the MEC and the MMRC. He is an *ex officio* member of MCC. He deputizes for the Dean as and when necessary.

Modular Course Co-ordinator
The Modular Course Co-ordinator takes particular responsibility for:

- meeting approved admissions targets by field;
- liaison with staff and students, for example to advise on the regulations;
- slotting and timetabling;
- liaison with student services; and
- student progress in Stage I.

In furtherance of these aims he acts as Chair of the Modular Admissions Committee (MAC) and Careers Liaison Group and Stage I Examinations Committee. He is an *ex officio* member of MCC, MEC and MMRC. He is required to deputize for the Assistant Dean and Dean as and when necessary.

The eleven senior tutors are each seconded from a department to the Modular Course for a third of their time. Their responsibilities for admissions and counselling in Stage I of the Course are set out in Chapter 4. Care is taken in allocation of responsibility for fields that each is concerned with at least one field of study outside his or her home department. In adjusting the Polytechnic's staffing establishment to account for their central contribution a formula is used which balances the contribution of numbers of staff from each department (0.3) – i.e. the 'payment' – against the central service each department receives (in terms of students admitted and counselled in Stage I) – the 'bill'.

Other individual contributions to Modular Course management are all departmentally based. The following extracts from the current Modular Course *Staff Guide* indicate the key expectations of the two most important links in the chain: Heads of Department and Field Chairs.

Responsibilities of Heads of Departments

- The Head of Department is responsible for providing the resources to teach modules in the way that has been agreed. This will require:
  1. an agreed departmental policy on the sizes of teaching groups

appropriate for the different modes of teaching used in different subject areas and the provision of sufficient staff hours for the modules to be taught in these ways;

2. consultation where appropriate with the Field Chairs and/or other staff over the timetabling of modules and the provision of specialist rooms and equipment at the times required for particular modules; and
3. a continuing review of staff and student FTEs so that appropriate admission targets can be agreed with the Dean, other Heads of Departments and the appropriate Faculty Dean, and so that Field Chairs can be properly advised on the implications of student transfers and exchanges.

- The Head of Department is responsible for ensuring that all personal tutors, module leaders and Field Chairs in his or her department are aware of their own responsibilities and must be prepared to make arrangements for someone to act in lieu of any of these staff who, for example, are not available to students or registry staff.
- The Head of Department or his or her nominee is responsible for informing the Registry of any limit on the numbers of tutees a particular member of his or her staff should be allocated, and of arrangements to be made for personal tutor responsibilities when a member of staff leaves or is granted leave of absence.
- The Head of Department is responsible for nominating persons with whom module leaders should clear their examination papers and mark sheets and for ensuring that associated deadlines are met.
- The Head of Department is responsible for the annual review of the fields sponsored by his or her department.
- The Head of Department is responsible for ensuring that proposals for change are considered by appropriate field and/or subject committees and the departmental board and in particular that resource implications are identified.
- The Head of Department in consultation with the Dean is responsible for recommending appropriate action in cases of suspected cheating or plagiarism by students.
- The Head of Department is responsible for seeing that access to projects and dissertations is as agreed with students.
- The Head of Department is responsible for discussing and agreeing with the Dean the withdrawal of modules from the course as a result of low recruitment or resource problems. In each case students registered for the module in the current or a future term must be counselled about and registered for viable alternative programmes.

Responsibilities of Field Chairs

There is a Field Chair for each approved field whose terms of reference are set out below. Some of these responsibilities are defined more specifically, but the way they are carried out will vary from field to field depending on the involvement of the appropriate Head of Department, the presence of Principal Lecturers with defined responsibilities, and whether there are several fields with common modules.

- to ensure with the appropriate Head of Department that the teaching and assessment of field modules is consistent with the content of the course as validated;
- to represent where appropriate the field in any external discussion with the CNAA, or other bodies;
- to test and report on the suitability of applications requiring dispensation;
- to counsel field students on their choice of modules;
- to arrange and chair, where appropriate, the subject examination committee;
- to arrange for field students to receive careers advice;
- to arrange and chair regular meetings of the field committee and to produce minutes;
- to discuss with the Co-ordinator, Assistant Dean and/or Dean any proposed changes in syllabuses, modules, slotting or field rules; and
- to represent the field on the MCC.

Terms of office
The Field Chair is elected for a two-year period by the staff members of the field committee and is eligible for re-election.

Beyond these defined central roles the Course has clear expectations of all permanent members of staff who contribute to it as module leaders and as personal tutors. In the former role they take responsibility for the design, teaching and assessment of individual modules (see Chapter 5). As the latter they counsel students about their programmes, confirm their specific choices, and take a direct interest in each student's personal and academic progress (see Chapter 4).

Groups

The committee structure set out in Figure 3.2 represents the collective decision-making framework of the Modular Course. As is clear from this diagram the main policy-making body is the MCC. Its more precise relationship to the substructure, in particular to MMRC and the field committees, is set out below.

Modular Course Committee (MCC)

- Membership
  Dean (chair)
  Assistant Dean

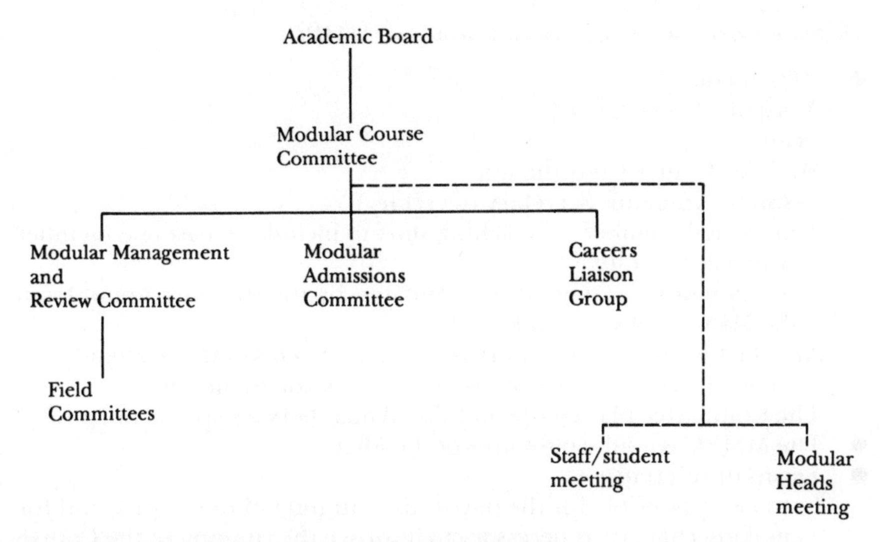

Figure 3.2 Modular Course Committee structure

Modular Course Co-ordinator
Field Chairs
Heads of Departments offering fields within the Modular Course
Senior tutors
Senior Administrative Assistant (secretary)
Examinations officer
Library representative
Four students' union representatives
Any member of MMRC not otherwise qualified.

- Terms of Reference
  The MCC is responsible for the coherence and educational effectiveness of the course, and for policy making, and reports to the Academic Board.
- Terms of office
  Field Chairs: elected by field committee for two years, renewable.
  Senior tutors: appointed for two years, renewable.
  Students: elected by the students' union at the start of the academic year.
- Meetings
  The MCC normally meets twice (in weeks 5 and 8) each term. The first meeting is termed a 'business meeting' and the agenda comprises such items as require decision for the day-to-day running of the course. The second meeting is termed a 'development meeting' and the agenda comprises items relating to the future policy and development of the Course. The 'development meeting' receives the annual review of the course from MMRC (see page 40) in Term 2 of each academic year.

Modular Management and Review Committee (MMRC)

- Membership
  Assistant Dean (chair)
  Dean
  Modular Course Co-ordinator
  Assistant Academic Secretary (secretary)
  Ten elected members of teaching staff to include at least one member from each faculty
  Two student representatives nominated by the students' union from the MCC representatives
  Any Field Chair (not otherwise qualified) choosing to attend and having two days' notice of his or her intention to the Chair.
  The Committee may co-opt additional members for special purposes.
- The MMRC is a sub-committee of the MCC.
- Terms of reference
  1. to be responsible for the day-to-day running of the Course and for making changes as necessary to improve the running of the Course within overall policy set by MCC;
  2. to set up *ad hoc* committees to examine certain aspects of the Course and supervise certain aspects of management;
  3. to review the regulations as required;
  4. to nominate members from the Course as a whole to Faculty Review Groups (FRG) when they are considering specific changes to fields, annual field reviews and new field proposals;
  5. to propose changes of policy to the MCC;
  6. to prepare briefing materials on issues referred to it by MCC or other bodies for the information of MCC;
  7. to provide the annual Course review for MCC; and
  8. to make recommendations on external examiner nominations to the ASC.
- Papers
  1. MMRC agendas are circulated to all Heads of Department and Field Chairs simultaneously with circulation to the Committee.
  2. MMRC minutes, once approved, are circulated to all members of MCC.
- The Committee normally meets fortnightly for the first eight weeks each term (weeks 2, 4, 6, 8) and may meet on a fifth occasion (in week 10).

Field committees

- Membership
  Field Chairs
  Module leaders of modules that are acceptable or compulsory for the field

Appropriate Heads of Departments (*ex officio*)
Student representatives
The committee may co-opt additional members for special purposes.

- Terms of reference

  The field committee is responsible for:

  1. administering, teaching and examining the field;
  2. making appropriate recommendations to the MCC for addition, deletion, or changes in modules or field rules; and
  3. nominating external examiners to MMRC and the relevant FRG.

- Terms of office

  Field Chairs are elected from the teaching staff members of the committee by the staff members. The office is held for two years from September and Field Chairs are eligible for re-election.

Admissions Committee (MAC)

- Membership

  Modular Course Co-ordinator (chair)
  Senior tutors
  Modular Course Admissions Officer (secretary)

- Terms of reference

  1. to agree the number of offers conditional upon A level performance that should be made to meet the admission targets for the fields and the Modular Course;
  2. to consider applications for places on the Course and to make offers, invite for interview or take other action as appropriate;
  3. to deal with entries entailing exemption from part of the Course on behalf of the Polytechnic Dispensations Committee, to set appropriate conditions for such entries and to report them to the Dispensations Committee; and
  4. to consider for approval any other exemptions proposed by Field Chairs. (This is not necessary for modules for which the grounds for exemption are specified in the module description.)

- Term of office

  Senior tutors are appointed by the Dean on the recommendation of the Heads of Department concerned. Appointment is normally from 1 January for two years, and may be renewed.

Careers Liaison Group

- Membership

  Modular Course Co-ordinator (chair)
  Head of Student Services (convenor)
  Careers counsellors from student services department
  Nominated representatives of fields or groups of cognate fields.

- Function
  1. to discuss relevant information and identify activities organized by student services of which academic staff and students should be aware;
  2. to provide a channel of communication between academic staff and the careers counsellors from student services;
  3. the group meets at least once a term, usually in week 2.

Heads of Departments meeting
The Dean and Co-ordinator meet each term (usually in week 11) with Heads of Departments contributing to the Modular Course to discuss matters of mutual concern. The notes of their meetings are presented to MCC.

Student/staff open meeting
An open meeting of students and staff is held in the first two terms of each year (usually in week 7), principally for students to express views and opinions to the Dean and staff. In the third term a meeting is held between students and the Dean or Assistant Dean and Co-ordinator alone, following an agenda set in consultation with the Students' Union. The notes of these meetings are presented to MCC.

Forward planning

There are several areas in which the continued health of the Course depends upon a collaborative response which must necessarily be more than the aggregate of views of individual fields and departments. A key example is the forward planning of student numbers. Here care has to be taken to establish the effect of decisions relating initially to one area across the Course and to fix, relatively early in the year, firm targets for student recruitment in order to meet the Polytechnic's plans for individual departments and schools. Further complications arise from the lack of fit of this planning process with the form of the overall NAB plan for the Polytechnic (as set out above in Chapter 2) and from such open-ended policies as the Polytechnic's commitment to expand overseas student numbers to 5 per cent of first year intake. The solutions are found each year through a judicious mixture of strong quantitative forecasting, relying on the student management system (see Chapter 7) and sensible qualitative judgement. The tables below indicate some major steps in the process of planning for 1988–89.

Table 3.2 gives the NAB plan as finally negotiated for Oxford Polytechnic. A first task accomplished during term 1 is to establish the current FTEs by school and department, fixed by a census on 1 November and to use both this information and the proposed NAB support to fix school and department targets for 1988–89. Table 3.3 gives this analysis for 1 November 1987. The academic staffing establishment included in the

Table 3.2 Student number targets (FTEs) 1988–89, 2 December 1987 as proposed by NAB board

LEA: Oxfordshire 931
Institution: Oxford Polytechnic 8714

Programme	*1986–87 monitoring*		*1987–88 targets*		*1988–89 targets*	
	FY	*AY*	*FY*	*AY*	*FY*	*AY*
01 Engineering	134	305	140	290	135	290
02 Other Technology	57	132	45	76	45	76
03 Construction	347	921	306	856	316	861
04 Science	190	456	181	514	190	523
05 Environmental Science	264	395	93	306	93	306
06 Agriculture	0	0	0	0	0	0
07 Health	56	120	16	16	16	16
08 Mathematics	95	220	166	309	179	352
09 Business Management	344	801	407	802	421	841
10 Catering	89	222	92	209	90	217
11 Other Professions						
12 Social Studies	84	222	92	209	90	217
13 Languages	57	181	76	234	75	243
14 Humanities	62	176	76	193	74	191
15 Art	0	1	0	0	0	0
16 Design	0	0	49	98	49	98
17 Performance Art	0	0	8	18	8	18
18 Initial Teacher Training	137	279	165	360	173	329
19 Inservice Education Teaching	270	380	176	220	161	235
Total	2,186	4,830	2,073	4,716	2,105	4,834
FTE maxima:						
First and higher degree:	1,342	3,587	1,410	3,661	1,399	3,668
Full-time and sandwich:	1,862	4,230	1,864	4,261	1,852	4,277
Mature Participants %:			40		40	

budget for that year is then fixed by the Planning Committee and Academic Board.

A proportion of these FTEs is then established as modular and an analysis of the likely effect of rolling forward current Modular Course student numbers is undertaken, including 'self-correcting' assumptions about the pattern of student choice (based on a year-on-year analysis of change) plus planned new initiatives. The outcome is set out in Table 3.4. Moving from left to right this establishes the proportion of each student registered on a field which is likely to contribute to each subject area's FTE in each stage (b,c), sets this against current FTEs (agreed in the 1 November census – after some major 'servicing' arrangements are allowed for; for

Table 3.3 Modular Course module registrations 1 January 1987

Student/staff ratios analysis by subject groups of student module registrations
Given average number of module credits per full-time student: 09.48
Each basic module credit registration is counted as ⅙ of a credit
Modes included: ALL

Subject group	*SSR*	*Module credits* Term 1	*Term 2*	*Term 3*	*Total credit*
04 Music	09.00	82.3	79.1	101.1	262.5
05 Unattached	12.00	0.0	40.8	70.8	111.7
21 French	12.00	255.6	223.4	447.3	926.3
22 German	12.00	188.8	156.5	261.2	606.4
23 English	15.00	439.8	400.2	373.7	1213.7
24 History	15.00	367.3	300.0	377.0	1044.3
25 History of Art	14.00	268.2	202.5	208.0	678.7
26 Geography	11.00	179.5	254.4	254.5	688.4
27 Italian	12.00	44.4	40.0	40.6	125.0
28 Spanish	12.00	48.9	43.9	42.2	135.0
31 Catering	12.00	440.8	487.4	901.6	1829.8
32 Food Science & Nutrition	10.00	125.0	183.0	128.0	436.0
42 Cartography	12.00	157.7	123.4	220.3	501.4
44 Construction	11.00	19.0	36.0	28.0	83.0
52 Publishing	12.00	239.8	208.8	308.2	756.8
57 Visual Studies	10.00	289.6	300.7	331.9	922.2
61 Education	11.00	310.3	240.2	306.3	856.8
62 Education	11.00	413.0	399.2	367.2	1179.3
66 BEd in-service	12.00	35.0	32.0	59.0	126.0
70 Business Administration	12.00	0.0	0.0	78.3	78.3
71 Accounting	14.00	261.2	246.0	299.5	806.7
72 Tourism	13.50	151.7	130.8	100.0	382.5
73 Anthropology	11.00	212.5	222.3	235.2	670.0
75 Economics	14.00	346.3	289.3	235.2	870.8
76 Politics	15.00	256.2	190.5	238.0	684.7
77 Psychology	11.00	341.2	330.0	238.8	910.0
78 Sociology	14.00	173.7	210.3	169.0	553.0
79 Law	14.00	379.7	295.3	341.0	1016.0
81 Biology	10.00	437.3	335.0	597.5	1369.8
82 Biology	11.00	292.7	401.2	393.3	1087.2
83 Geology	11.00	262.3	315.5	477.0	1054.8
84 Maths statistics and computing	14.00	345.8	250.8	124.2	720.8
85 Physical sciences	11.00	293.8	210.8	312.2	816.8
86 Mathematical studies	14.00	102.7	184.8	241.0	528.5
87 Computer studies	12.00	522.3	381.5	375.8	1279.7
89 Microelectronics	11.00	45.8	2.0	0.0	47.8
91 Planning	10.00	0.0	28.0	28.0	56.0
92 Architecture	12.00	4.2	5.8	8.3	18.3

example a high proportion of Stage I accounting is taught by staff from the catering fields) (a), and adds the likely effect of recruitment targets and combinations for September 1988 (ij), to produce an FTE prediction for the next year (m). These are then recast into a departmental format and presented for discussion to a key meeting of the central academic team and the Heads of Departments contributing to the Modular Course at the end of term 1.

At this meeting there is considerable scope for adjustment, but the process would clearly fail if there were not also an understanding of the need for a consensual solution to emerge, and within a very tight timescale. Usually the proposed adjustments are checked outside the meeting by the central team and recirculated for quick comment before publication in a final form in which they can be used both by the Modular Admissions Committee (MAC) to control enrolment and by faculties and departments as the base for resource management. In December 1987 this adjustment process took a further week and the final outcome is as recorded in Table 3.5.

The fragility of this process is apparent to all who participate in it. It is crucially dependent upon accurate forecasting and the ability of the MAC to deliver the right student numbers in the right field combinations. Experience has shown that the forecasting system is fairly robust in years in which there are no major changes in the shape of the Course offerings. In 1987–88 this was not the case and we failed to predict a dramatic upturn in School of Business FTEs, principally as a consequence of students electing to take tourism modules as an extra-field choice. (Compare the targets with the actuals in Table 3.5.) The Course as a whole also over-recruited significantly. 1988–89 is a similarly uncertain affair with the new dimensions of planning, business administration, physics and chemistry. In these circumstances it was no accident that planning and resourcing were high on the agenda for internal debate.

A year of policy-making

As indicated above, the engine room of policy-making and analysis is the MMRC, with its largely elected staff membership. Its role is to formulate options and propose solutions to the MCC, which is ultimately responsible to the Academic Board for the Course and its development. It also plays a key role in the review and validation system. The analysis below of a year of MMRC activity illustrates not only the nature of live business but also the volume and responsibility of the work it undertakes.

MMRC business 1987–88

The Committee met, as usual, at least five times in each term, often jointly with a FRG. The business outlined below has been collected under three

Table 3.4 Module FTE predictions for 1988–89

	Fractions		*Students*		*1987 Current FTEs*			*Targets*		*1988 Predicted FTEs*		
FIELD *a*	*F1* *b*	*F2* *c*	*S1* *d*	*S2* *e*	*CALC* *f*	*ADJ* *g*	*ACTUAL* *h*	*S1* *i*	*S2* *j*	*CALC* *k*	*ADJ* *l*	*FTE* *m*
AC	0.04	0.45	39	83	38.91	38.91	55.6	30	86	39.90	39.90	52.13
AN	0.25	0.40	48	92	48.80	55.36	70.7	45	89	46.85	52.43	61.24
BI	0.33	0.50	55	105	70.65	0.00	0.0	55	116	76.15	0.00	0.00
BA	0.25	0.45	1	0	0.25	5.83333	8.3	25	0	6.25	16.25	21.14
EB	0.50	0.75	19	42	41.00	0.00	0.0	20	43	42.25	0.00	0.00
ES	0.25	0.40	43	78	41.95	0.00	0.0	35	79	40.35	0.00	0.00
CB	0.5	0.9	20	0	10.00	0.00	0.0	25	20	30.50	0.00	0.00
HB	0.50	0.70	32	70	65.00	240.83	259.2	25	63	56.60	258.78	254.68
CT	0.25	0.50	51	66	45.75	74.44	61.7	45	70	46.25	75.97	57.58
CA	0.33	0.50	24	44	29.92	0.00	0.0	25	44	30.25	0.00	0.00
CD	0.42	1.00	76	108	139.92	169.84	189.6	75	107	138.50	168.75	172.26
CH	0.25	0.5	1	0	0.25	49.0567	43.1	30	0	7.5	49.15	39.49
FN	0.08	0.40	26	79	33.76	40.76	46.0	32	76	33.06	39.36	40.62
EC	0.33	0.50	44	67	48.02	62.72	91.9	27	75	46.41	57.77	77.39
NF	0.42	0.60	49	72	63.78	0.00	0.0	51	78	68.22	0.00	0.00
ED	0.25	0.50	49	109	66.75	0.00	0.0	40	110	65.00	0.00	0.00
JM	0.42	0.60	35	73	58.50	189.03	214.8	37	66	55.14	188.36	195.72
EN	0.25	0.50	87	175	109.25	117.87	128.0	80	178	109.00	116.08	115.27
FS	0.42	0.55	37	42	38.64	0.00	0.0	29	45	36.93	0.00	0.00
FL	0.42	0.55	31	32	30.62	93.51	74.9	27	33	29.49	114.42	83.80

GG	0.33	0.60	58	101	79.74	79.74	72.6	55	105	81.15	81.15	67.56
EA	0.33	0.60	40	86	64.80	0.00	0.0	46	80	63.18	0.00	0.00
GL	0.33	0.50	19	18	15.27	85.72	111.3	20	18	15.60	84.08	99.82
DS	0.42	0.55	23	33	27.81	0.00	0.0	24	30	26.58	0.00	0.00
DL	0.42	0.55	16	29	22.67	59.26	55.8	18	27	22.41	66.39	57.17
LB	0.22	0.26	67	2	15.26	15.26	27.4	40	62	24.92	24.92	40.91
HA	0.33	0.50	50	92	62.50	62.50	71.6	42	91	59.36	59.36	62.18
HI	0.33	0.50	79	149	100.57	108.07	110.2	65	150	96.45	103.95	96.92
LW	0.33	0.66	38	83	67.32	81.62	107.2	27	82	63.03	79.04	94.92
CO	0.33	0.50	71	106	76.43	0.00	0.0	60	111	75.30	0.00	0.00
MA	0.42	0.50	42	66	50.64	224.60	266.7	36	71	50.62	234.06	254.14
ME	0.08	0.5	41	3	4.78	4.94667	5	30	38	21.4	26.4	24.40
MS	0.50	0.50	27	34	30.50	30.50	27.7	25	33	29.00	29.00	24.08
PY	0.50	0.80	13	36	35.30	0.00	0.0	0	30	24.00	0.00	0.00
PH	0.42	0.50	19	31	23.48	0.00	0.0	15	27	19.80	0.00	0.00
PC	0.25	0.5	1	0	0.25	42.5567	43.1	30	0	7.5	42.817	39.65
PL	0.9	0.9	1	0	0.9	0	0	50	0	45	0	0.00
PN	0.33	0.5	1	0	0.33	8.56333	7.8	20	0	6.6	56.6	47.14
PO	0.33	0.50	34	92	57.22	69.57	72.2	27	83	50.41	63.59	60.35
PS	0.16	0.40	37	111	50.32	50.32	96.0	37	105	47.92	47.92	83.60
PB	0.33	0.50	49	98	65.17	0.00	0.0	40	98	62.20	0.00	0.00
PU	0.42	0.75	0	3	2.25	67.42	79.8	0	0	0.00	62.20	67.32
SO	0.33	0.50	41	69	48.03	54.61	58.3	37	72	48.21	53.63	52.35
TS	0.25	0.5	44	2	12	24.75	40.3	30	46	30.5	45.083	67.12
VS	0.33	0.5	25	83	49.75	49.75	97.3	30	77	48.40	48.40	86.56
UNATT							11.8					10.62
			891.5	1505	1945	2257.9	2606	929.0	1568.5	2024.1	2385.8	2497.5

Table 3.5 Modular Course FTEs by department and recruitment by field, 1988–89

Department/school	*Actual FTEs 1987–88 (targets)*	*1988–89 predicted FTEs*		
		By school or dept	*By subject*	*Field targets**
Biological and molecular sciences	348	367	BIOL 279	BI 55
	(343)		CH 44	EB 20
			FN 44	ES 35
				HB 25
				CB 25
				CH 30
				FN 32
Hotel and catering management	223	223		CA 25
	(220)			CD 75
Civil engineering, building and cartography	62	63		CT 45
	(60)			
Visual arts, music and publishing	205	195		MS 25
	(200)			PB 40
				VS 30
Education	228	228		AE 88
	(222)			ED 40
				LS ?
Engineering	48	70	ME 26	ME 30
	(50)		PC 44	PI 30
				PH 15
Geology	111	109		EA 46
	(105)			GL 20
Humanities	311	300		EN 80
	(310)			HH 65
				HA 42
Business	375	433	LPE 259	LW 27
	(307)		BA 50	PO 27
			TS 75	EC 27
			AC 57	BA 26
				TS 30
				AC 30
Computing and mathematical sciences	267	278		CO 60
	(240)			MA 36
Modern languages	189	237		FL 27
	(188)			FS 29
				DL 18
				DS 24
				LB 40
Social studies	225	215		AN 45
	(215)			PS 37
				SO 37
Planning	8	52		PL 50
	(6)			PN 20
Geography	73	74		GG 55
	(73)			
Unattached	12	12		
	(7)			
Totals	2,658	2,855		923
	(2,552)			

*Initials represent field codes

not mutually exclusive headings: validation, review and policy. Policy items normally resulted in a recommendation to MCC, taken at the next development meeting of that Committee. Validation recommendations normally went straight to the Polytechnic's ASC. External examiner approvals were also recommended to ASC.

Term 1

	Validation	*Review*	*Policy*
Meeting 1	Modifications to applied education to meet CATE (Council for Accreditation of Teacher Education) criteria. Recommendation to ASC	Discussion of CNAA letter of approval for languages for business Establishment of working party to review slotting Review of *Staff Guide* guidelines on double marking (requested by ASC)	Discussion of change of regulation for counting of double projects: recommendation to MCC Establishment of working party to review examination arrangements Preliminary discussion of problem of large unanticipated enrolments on some modules
	(+1 external examiner approval)		
Meeting 2	New prerequisite for module, 'Basic keyboard and information skills': recommendation to ASC		Discussion of possible development of field in public relations Agreement to establish course planning committee for new double field in applied geology: recommendation to ASC
	(+1 external examiner approval)		
Meeting 3 (Joint with arts and education FRG)	Initial discussion of proposed change to history field	Review of Education and BEd for qualified teacher fields	

	Validation	*Review*	*Policy*
Meeting 4 (Joint with Technology and life sciences FRGs)	Discussion of proposed single fields in physics and chemistry		
Meeting 5	Discussion of proposed module changes in biology: recommendation to ASC	Review of basic unattached modules ('Women's experience in Britain today' and 'The threat of nuclear war') Report of working party on slotting	Proposed changes to regulation on entry with advanced standing Discussion of student numbers on tourism modules
Term 2			
Meeting 1	Proposed new sociology module 'International dimensions of development policies': recommendation to ASC		Discussion of implications of accreditation (all FRG Chairs invited) Further discussion of open access to modules: recommendation to MCC Report of working party on examining
Meeting 2	Change in assessment of geography modules: recommendation to ASC Proposed exclusion rules for planning basic modules: referred (+4 external examiner nominations)	Discussion of CNAA letter of approval and draft reply on tourism Receipt and discussion of *Annual Review 1986–87*: recommendation to MCC	Preliminary discussion of possible new qualification in vocational training

	Validation	*Review*	*Policy*
Meeting 3	Changes in assessment of modules. Specific proposal in psychology and agreement on procedures in the light of working party report on examining: recommendation to ASC	Response to CNAA on languages for business and tourism: recommendation to ASC	Counting rule for projects: arrangements for implementing MCC decision
	(+3 external examiner nominations)		
Meeting 4			Agreement to set up course planning committee for new double field in computing Agreement on final form of regulations on project counting and admissions with exemption
	(+2 external examiner nominations)		
Meeting 5 (Joint with life sciences FRG)	Preliminary discussion of submission for nursing and midwifery double fields	Cell biology: response to CNAA conditions: recommendation to ASC	
Term 3			
Meeting 1	Proposed changes to French fields: recommendation to ASC	New procedures for considering exemptions from period of residence abroad for modern languages	Discussion of proposed change to deadline for student withdrawal from modules: recommendation to ASC
	(+2 external examiner nominations)		

	Validation	*Review*	*Policy*
Meeting 2	Changes to English field: recommendation to ASC Proposed new modules in history: recommendation to ASC Changes to publishing field: recommendation to ASC Changes to law field: recommendation to ASC	Tourism: response to CNAA conditions: recommendation to ASC	Establishment of Named Polytechnic Diploma in Advanced Studies: recommendation to MCC
Meeting 3	Changes to education field: recommendation to ASC Changes to Stage I mathematics modules: recommendation to ASC (+4 external examiner nominations)	Review of physical sciences single field in the light of approval of physics and chemistry	Discussion of student success rates in Stage I as a consequence of changes to regulations in 1984–85
Meeting 4 (Joint with arts and education FRG)	Proposed changes to BEd in-service field: referred for further discussion Proposed changes to applied education field: recommendation to ASC Various changes in assessment in response to working party on examining: recommendation to ASC		Discussion of role of personal tutor in preparation for 'feedback' seminar

	Validation	*Review*	*Policy*
Meeting 5	Proposed changes to politics, visual studies and computer studies field	Review of exclusion rules required by introduction of planning fields	Preliminary discussion of MMRC workload
		Agreement on actions required by 1986–87 onward review	
		Establishment of timetable for major reviews of Modular Course fields 1988–93	

What this summary reveals primarily is the volume and range of work undertaken by a largely elected body of staff and students (it disguises a further aspect of their role as participants in FRG-led review events). It also points to the creative tension between the needs for *maintaining* the course (trouble-shooting – as in the 'open access' and 'examining' discussion – as well as fine-tuning the syllabuses offered by fields through the validation system), and those of creatively *developing* it (in discussions of new fields, new awards and review of basic principles). Against this background we can return to the issues outlined at the head of the chapter.

Academic and resource planning

As illustrated in the section on forward planning, neither type of planning can proceed in ignorance of the other. When they fail to be successfully coordinated, as in the episode of tourism enrolments, considerable problems emerge quickly.

Academic and administrative control

It is important that MMRC establishes for the Course management a clear set of academic priorities. While it is true that no major course development has been held up by administrative constraints, the range of decisions with administrative effects (deadlines, counting rules, etc., as well as the information base about approved fields) imposes a permanent culture of change on the administration of the Course.

Briefing and training

MMRC and MCC decisions can be sensibly implemented only with the knowledge and good will of staff. The lines of communications, between the two committees, and from the MCC to field committees, is vital. Periodic initiatives, such as the termly feedback seminars, the publication of in-house journals such as *Teaching News* (now incorporating the Modular Course's own broadsheet *Feedback*), and the regular issue of updates to the definitive *Staff Guide* all play a part. Systematic training, beyond a small contribution to the induction course for all new staff, has not yet been established.

Speed and flexibility of response

The data above do not give much direct information about timescale. Fortnightly meetings of MMRC obviously contribute to quick response, as do the Polytechnic's new responsibilities under accreditation. However, each proposal emerges here only after an important 'prehistory' of field committee, departmental board (and occasionally subject committee) consideration (see Figure 2.5: subject committees exist for some subjects which contribute to both modular and non-modular courses in the Polytechnic, for example sociology). The development of 'fast track' procedures, such as those referred to here for assessment changes is indicative of mounting concern.

Centre and periphery

This chapter has necessarily concentrated on activity at the centre and, in doing so, demonstrated how much any local initiative depends eventually on Course-wide endorsement. Speed of response, confidence that the centre will pick up local concerns, and that centralized constraints will be kept to a minimum are all important in maintaining good morale and a positive outlook by fields.

4

Admissions and Counselling

John Brooks

Admissions

General principles

The general aim of admissions work is to fill the Course with well motivated students who are likely to benefit from it. Applicants' ability to benefit is assessed flexibly and credit is given for a variety of qualifications beyond conventional O/GCSE and A level successes. This is particularly true in encouraging the admission of mature students, among whom local applicants are viewed especially sympathetically. A good social 'mix' is sought among British students and offers are commonly made to applicants both from the EEC countries and other parts of the world. Applicants with appropriate previous study and/or experience are given exemption from part of the Course and students wishing to transfer from another institution are welcomed, subject to numbers. By offering a range of study modes and routes, including part-time associate status, individuals are encouraged to use the Course for a variety of personal purposes.

Staffing and organization

Ultimate responsibility for admissions lies with the Modular Course Co-ordinator, an academic working essentially full-time in a management role but retaining some involvement in teaching. Admissions matters are the largest but by no means only aspect of his responsibilities. He works with a team of senior tutors whose number has increased broadly in line with the growth of the Course and of its popularity. There are currently eleven. Each may handle up to 2,000 applications from the Polytechnics Central Admissions System (PCAS) in a year and is responsible for counselling a proportion of Stage I students. The size of these responsibilities is recognized by 0.3 remission of teaching. Tutors are normally put forward by each department or school contributing to the Course and serve for two years. The Co-ordinator and tutors meet collectively as the MAC twice termly and as a Stage I examinations committee termly. The detailed administration of applications is handled by a full-time Modular Admissions Officer who draws on clerical support from within the Registry and who makes substantial use of the central computer services. Staffing is temporarily expanded by the employment of students from the Course, most notably from late July to late September, but also at other periods of particular pressure during the year.

Targets and offers

The number of students to be recruited for each field in the next academic year (the target) is fixed during the first term of the current year. The targets relate to first-year degree applications, very largely received through PCAS. Some such applications for the BEd degree are received from the Central Register and Clearing House (CRCH) but, unusually, BEd candidates may alternatively apply to Oxford through PCAS. These figures are agreed by consultation with departments and schools on the basis of FTEs known to be generated by the present students and projections for the following year. In practice many targets remain unchanged for the following year and alterations are normally within the range of 10 per cent up or down. There is a wide variation in targets, those for 1988 entry varying from 88 for applied education and 80 for English literature to 18 for German language and literature and 15 for physical sciences (single field). Many of the fields most popular with applicants have targets of modest size. The 1988 targets of 26 for business administration, 27 for law and 37 for psychology meant ratios of applications per place between September and mid-December 1987 of approximately 75 : 1, 60 : 1 and 55 : 1 respectively.

Also the subject of discussion and refinement during the first term are the number of offers to be made for each field to reach the target (the offer ratio) and the A level grades normally required for entry. Adjustments are considered in the light of experience of the immediately preceding cycle. The ratios in use for 1988 entry range widely, reflecting knowledge of the likely number of applicants and the probable take up by those to whom offers are made. Examples across the range are environmental biology (double field) and mathematics, 13 : 1, English and history, 9 : 1, and publishing and visual studies 4 : 1. High offer ratios often reflect the relative difficulty of meeting the targets for a small number of fields. However, they are also set high for a number of fields which are very popular with applicants; used in conjunction with offer conditions which are also high, as a controlling device, they enable offers to be made to a more reasonable proportion of the total applicants (although for 1988 entry 70 per cent or more of all applicants were rejected for twelve very popular fields).

Normal conditional offers broadly reflect the supply of places and student demand for them. Recent adjustments have usually been slightly upwards, necessitated by the wish to make offers to more applicants, as in the cases of law and psychology for which grades of BCC were required in 1987–88. Most applications are for two single fields, the normal requirements for which are often different, in which case the conditions are averaged. For example, an offer for law (BCC) and planning (DDD) would be set at CCD. It follows that applicants wishing to combine two popular fields are set stiffer conditions (e.g. psychology with English BCC), while lower grades are asked of those choosing a popular and a less popular one (e.g. psychology with chemistry DDD). Offers are always expressed as two or three grades rather than a total of points to retain a degree of control at the confirmation stage. Passes, sometimes at specified grades, in individual subjects usually reflect the particular entry requirements of a field. Senior tutors may make offers on conditions differing from the norm to take account of an applicant's special circumstances, but it is stressed that this should be an exceptional practice in case targets be overshot. The A level grades normally required are published in the *Prospectus*, as are fields' specific entry requirements. The introduction of new fields into the system is problematic as offer ratios and the grades normally required have to be set initially without the guidance of direct previous experience.

There is some intervention in the matter of field combinations. A field which wishes to attempt to ensure that some offers are made in combination with one or more other fields sees to it at the beginning of the cycle that 'reserved offers' are set aside. Such reserves are most commonly made by fields which, for various reasons such as the general practice of interviewing, tend to fill more slowly (e.g. applied education and visual studies), and with other fields which tend to fill quickly (e.g. psychology). Reserved offers are subtracted from the maximum offers to be made and then added in by the Co-ordinator as used up. Reserved offer combinations which prove irrelevant, for the opposing reasons that applications for them are abundant or virtually non-existent, are dispensed with. Attempts have also been made in the past to limit rather than protect the numbers of offers for particular combinations (e.g. English literature and history of art) to maintain a good spread of students.

PCAS applications and the monitoring of offers

The flow of PCAS forms, which by mid-December 1987 amounted to over 16,000, begins early in October. The forms of a number of applicants who have chosen only one single field are temporarily delayed while their choice of a second single field is ascertained. Forms are forwarded to senior tutors on the basis of their current 'ranking'. Table 4.1 shows that in use for 1988 entry.

The basic purpose of the ranking is to equalize the workload among tutors and to determine which one of them will make a decision on each

Table 4.1 Rankings used for 1988 entry

Senior Tutors (initials)	*Responsible for these fields*
1. R.B.	Education, applied education, music, visual studies
2. R.H.	Catering (single and double), tourism
3. I.L.	All languages fields
4. A.C.	Cartography, earth sciences, geology, physical sciences fields
5. S.W.	Geography, planning (single and double)
6. C.J H.	Law, politics
7. S.P.	All biology fields, food science
8. A.P.	Accounting, economics, business administration
9. W.T.	Computing, mathematics, microelectronics, and all overseas applicants
10. A.G.	Anthropology, psychology, sociology
11. S.W.	English, history, history of art, publishing

applicant. Workloads of applications and counselling are reviewed in the current year and adjustments made to the ranking for the following year, by moving tutors up or down, or by moving responsibilities for individual fields. Again, the introduction of new fields causes uncertainty.

Each tutor is responsible for a broadly related group of fields and automatically receives all applications for any double fields among them. However, the number of applications received by him/her for the single fields depends on the ranking. At the top R.B. receives all applications in which one of his four fields feature in the combination. In the middle C.J.H. receives applications for law and politics other than in combinations with single fields handled by the five tutors above her. At the bottom S.W. receives only applications for combinations among his four fields, all others having been directed to tutors above him. Because of the importance of a co-ordinated and experienced approach to applications from overseas all such, whatever the fields, are handled by W.T., whose workload of home applications is accordingly lower.

Decisions to offer or reject applicants are made by the senior tutors, in consultation with the relevant Field Chairs and/or Co-ordinator if necessary. Pressure of applications means that the use of interviews is exceptional, for most fields, being limited, for example, to mature students offering non-standard qualifications. However, all applicants under serious consideration for the fields of applied education, planning (double field) and visual studies are interviewed. The decision on each applicant is entered on the computer record, communicated to PCAS, and the form filed.

Table 4.2 Numbers accepted on an imaginary field in January of any year

Field	XX	
Target	30	
Maximum offers to be made	255	(i.e. 30 × offer ratio of 9=270, less 15 unused reserved offers)
Total offers made – raw	132	(i.e. a simple total of all offers)
Total offers made – derived	262	(see text below)
Conditional insurance	35	
Conditional firm	90	
Unconditional insurance	2	
Unconditional firm	5	
Rejected	700	
Withdrawn, cancelled, declined	75	
Outstanding interviews	3	
Other applications	5	
Total applications for field	915	

The cumulative effect of decision-making is communicated to tutors and the Co-ordinator by the Admissions Officer through the weekly issue of a computer print out known as a P11. This records, in column form, the following information for each field. The numbers given in Table 4.2 are for an imaginary field in, say, January.

The crucial column is the derived total of offers made which can be compared with the maximum number to be made as a measure of the field's fullness. Until late May this figure is calculated as follows: conditional insurance number+conditional firm number×2+unconditional insurance number+unconditional firm number×offer ratio; i.e. in the above case 35+180+2+45=262. As the 'derived offers' figure rises during the cycle it is automatically flagged with CARE at 75 per cent of the maximum to be made, STOP at 85 per cent and FULL at 100 per cent. Once a field reaches STOP all further proposed offers must be referred to the Co-ordinator, but in practice it is difficult to avoid overshoots in the more popular fields in the spring. For 1988 entry the method of calculating the derived figure has been changed as from the end of May when applicants have decided on their choices. Based on the experience that very few who opt to hold offers as insurance actually enrol, their numbers have been omitted from the later calculations. The derived total becomes the conditional firm number+the unconditional firm number×the offer ratio.

Given the fullness of most fields by the end of the first main phase of the PCAS cycle, the continuing applications procedure and late applications are relevant for only a minority of subjects, sometimes including new fields which suffer from uncertainty as their approval is pending.

The accurate hitting of targets in September owes much to the decisions made as to whether or not to confirm offers to those who have not quite met the conditions, a process strongly influenced by estimates (based on past experience) of the likely take up by those that have. These decisions in late August are made solely by the Co-ordinator, in the interests of the Course as a whole. Confirmations of offers which help 'weaker' fields are particularly important.

Regular print outs to monitor the situation change in format and name in late August from P11s to P13s. The latter charts the position field by field, recording the numbers of applicants in various categories, most importantly those with unconditional offers who have firmly accepted (UF) and those with unconditional offers whose decision is as yet unknown (U-). An estimated enrolment figure is calculated by the addition to the UF figure of the U- figure weighted according to previous experience. Thus in one field the U- figure may be given a value of 90 per cent as take up is known to be very strong, while in another it is given a value of 25 per cent since only 1 in 4 of such applicants is thought likely to accept. The validity of these take up assumptions is obviously affected by the passage of time from late August to late September and the firmness of the UF figures is known from experience to vary widely between fields; the estimated enrolment figures are therefore used only as an approximate guide to the Co-ordinator and senior tutors in deciding on the numbers of late offers to be made at special weekly meetings of the MAC in September.

Special classes of applicant

Like other institutions the Polytechnic has paid increased attention to the recruitment of overseas students in recent years. For the Modular Course one response has been the centralization of liaison and decision-making in a single senior tutor. A second response has been the setting of separate targets for each field, over and above those for home students. When first set for 1987 entry, targets were put at 10 per cent of the home figure for most fields and a blanket 15 : 1 offer ratio was adopted. The figures for 1988 entry have been revised in the light of experience, especially for the small number of fields for which demand is strong, such as accounting & finance, computer studies and law. On the one hand their targets have been increased by agreement with the relevant Heads of Departments/Schools, in one case to a figure as high as 66 per cent of the home target, and on the other hand some offer ratios have been varyingly revised down. However, it is recognized that decisions by overseas students to accept or reject offers are influenced by a more complex set of factors than for home students, and there remains uncertainty over the appropriate offer ratios to be applied. Detailed work is likely to be done on this aspect in the future.

The preceding paragraph relates exclusively to applicants who are both liable to pay overseas fees and who are seeking places on a degree course.

Students possessing one but not both characteristics are also welcomed onto the Course. A number of American students spend from one to three terms at the Polytechnic, sometimes as part of institutional exchange arrangements, sometimes on individual initiative. The modular nature of the Course facilitates credit transfers to and from the American system. Many EEC students are enrolled on the Course, usually either as one year exchange or private associate students; but a small number seek full degree programmes.

Non-degree applicants apply direct to the Polytechnic, as do students wishing to undertake degree programmes by part-time study. The extreme flexibility of the Course allows part-time students to change later to a full-time mode, just as it allows full-timers to drop down to part-time study. Decisions on part-time degree applicants are normally taken by senior tutors.

The Polytechnic Diploma in Advanced Study has proved an extremely valuable opportunity for certain types of applicant, frequently already possessing a first degree. The individually tailored package of at least 7 modules is ideal for those seeking professional re-orientation or updating in fields such as publishing, computer studies and cartography. Since the programmes are predominantly of advanced modules, admissions decisions are made by the relevant Field Chairs.

The Course is heavily used by part-time Associate students, mainly home students drawn from the local catchment area. Their reasons for study are various and include simple personal interest and skill acquisition in connection with present or possible future employment. Some in due course opt for a diploma programme. Particularly notable are those who use the Course itself as an access device. Prospective applicants, usually mature, are given the option of studying on the Course as associate students to demonstrate their potential as degree students. Asked to take normally from 2 to 4 modules (depending on their existing qualifications) they apply through PCAS and are made offers conditional on their performance as associates. All associate applicants apply direct to the Polytechnic on a simplified application form. Subject to establishing their ability to benefit, in which a reference plays an important part, they are admitted by the decision of the Co-ordinator. No limits are in force and such students simply join the others already on the modules of their choice. Most part-time students pay fees of one-twelfth of the full rate for each modular credit, up to a maximum of 6 in a year. For Oxfordshire residents who are registered unemployed or retired the fees are waived for part-time associate study, and from 1988–89 part-time degree students will pay only half (£24) the normal rate per credit.

It is conceivable that quotas may come into future use to encourage or limit the numbers of certain types of applicant. There is current concern to increase the numbers of mature students at the Polytechnic and, although the record of the Modular Course is good in this respect, quotas may be introduced. Also of current concern is the question of the social composi-

1. Schedule for Admission with Partial Exemption from Stage II

Regulation 12 on page 8 refers

A candidate for a degree or an honours degree who has been granted exemption from between one and two years of the course for BA and BSc candidates, or between one and three years of the course for BEd candidates, shall be required to pass (A) further module credits of which at least (B) must be acceptable for the field(s), within (C) terms. An honours candidate choosing to complete in a longer period than (D) terms shall not be permitted to take more then (E) module credits.

For an honours candidate the classification of the degree will be decided on the average of the (F) further module credits in which the candidate has been awarded the highest marks.

A students further module credits may include up to (G) basic non-acceptable module credits other than those for which exemption can be claimed. In addition, up to (G) acceptable basic module credits may be included.

BA and BSc Degree and Honours Degree candidates

Number of Module Credits to be passed		Maximum number of acceptable module credits		Maximum number of terms	Maximum number of terms of full-time study	Maximum number of module credits allowed if extended beyond D	HONOURS on best	Number of BASIC module credits
A		B		C	D	E	F	G
module credits								
for Honours	for Degree	for Honours	for Degree	Terms	Terms	Module credits	Module credits	Module credits
18	*16*	*16*	*14*	*15*	*6*	*21*	*18*	*2*
17	15	15	13	14	6	20	17	2
16	14	14	12	13	5	19	16	2
15	13	13	11	13	5	18	15	2
14	12	12	11	12	5	16	14	2
13	12	12	11	11	4	15	13	1
12	11	11	10	10	4	14	12	1
11	10	10	9	9	4	13	11	1
10	9	9	8	8	3	12	10	1
9	8	8	7	8	3	11	9	0

2. BEd Degree and Honours Degree candidates

Number of Module Credits to be added		Minimum number of acceptable module credits	Maximum number of terms	Normal number of terms of full-time study	Maximum number of module credits allowed if study extended beyond D	HONOURS on best	Number of BASIC module credits
A module credits		B	C	D	E	F	G
for Honours	for Degree	for Honours	Terms	Terms	Module credits	Module credits	Module credits
* 27	25	25	21	9	32	25	2
26	24	24	21	9	31	As column A less the number of S graded teaching practice module credits (to a maximum of 2)	2
25	23	23	21	8	30		2
24	22	22	20	8	28		2
23	21	21	19	8	27		2
22	20	20	18	7	26		2
21	19	19	18	7	25		2
20	19	19	17	7	24		2
19	18	18	16	6	23		2
18	17	17	15	6	21		2
17	16	16	14	6	20		2
16	15	15	13	5	19		2
15	14	14	13	5	18		2
14	13	13	12	5	17		2
13	12	12	11	4	15		1
12	11	11	10	4	14		1
11	10	10	9	4	13		1
10	9	9	8	3	12		1
** 9	8	8	8	3	11		0

* The figures in the lines marked * indicate the normal Stage II requirements.

** The figures in the lines marked ** indicate the Stage II requirements for a BA or BSc candidate with two years exemption, or a BEd candidate with three years exemption, from the whole course including Stage I.

Figure 4.1 Form M103, illustrating procedure for calculating exemptions

tion of the Polytechnic's student body, in particular the 'Sloane' image it has acquired. Such a dimension is not readily measured. The most accessible data which might serve as a yardstick is that on applicants' type of educational establishment. Analysis of this information is planned, an exercise which could lead to policies designed to limit offers to applicants from certain types of schools/colleges. Any such policy of quotas or limits would add to the complexities of making and monitoring offers, given the multiplicity of fields and combinations (some with reserved places).

Admissions with exemption

There is a steady flow of applications both from students with educational experience at an appropriate level in the past and from current students either at other institutions or on non-modular courses at Oxford. Although such applications are time-consuming, they make a significant contribution to the Course; in September 1987 60 students enrolled as direct entrants to Stage II.

Application is made direct to the Polytechnic and decisions on whether to offer a place and what exemptions are to be given are made by the relevant Field Chairs, using an interview if necessary. Exemptions agreed can be anywhere between the extremes of 1 or 2 basic modules to all but 9 advanced ones, depending on the amount and level of previous study and the closeness of its relationship with the new course. In effect the regulations are tailored to fit each individual. Examples of these arrangements, as recorded on the M103 forms used, are given in (Figures 4.2 and 4.3). In cases where the normal rules in Stage II do not apply the entries in Section 3 are made according to the illustrated schedule in Figure 4.2. All cases are formally approved by all senior tutors sitting collectively as the MAC, thus ensuring fairness and consistency of practice.

Phillip is 34 and has several qualifications in civil engineering and related areas. These provide the basis for exemptions from 5 basic modules shown in Section 4 – 3 in cartography and 1 each in planning and geology. No more exemptions are appropriate as his choice of degree subjects indicates a change of direction into completely new fields. Since his educational experience includes no work in anthropology or sociology his entry to Stage II is conditional on passing the 3 basic modules compulsory for each of those fields. He will study these as a part-time student in 1988–89 and enter Stage II in September 1989.

Nikola (Figure 4.3) is a straightforward case of a transfer from the first year of one degree course direct into the second year of another. The delay in starting the course until January was caused by her leaving an interval of a few months before approaching Oxford. Because there is a good general correspondence between courses she has been exempted from the 3 history compulsory basic modules and from 2 politics ones. It has not been possible to exempt her from a third politics module, 7605, but rather than delay her

OXFORD POLYTECHNIC MODULAR COURSE - Admission with Exemption Form M103

1 Stage II Entry Date ..September 1989.....[Month and Year] DCM minute number: 9 4 7

Surname......................... Forenames..Phillip..............................

Field(s)..AN/SO.. Date of Birth..14.10.1954.....

2 **Approval for admission** is XXXXXXXXXXXXXXXXagreed, on condition of passing: [**ring word * required**]

modules 7300, 7304, 7302, 7801, 7800, 7804.

and on the basis of the following **previous educational experience:**

Year(s)	Level	Subject	Institution
1976	ONC	Mining	Swansea College of H.E.
1979	OTC	Surveying etc.	" " "
1981	OTC	Civil Engineering	Lincolnshire Tech. College
1983	HTC	" "	N.E. London Polytechnic
1988	Diploma	Highway & Traffic Engineering	Middlesex Polytechnic

Other relevant experience...

English Language requirement satisfied? Y [Y or N] **CNAA** approval needed N [Y or N]

3 **Proposed/agreed Stage II programme is for** [Y] **Honours** [ ] **Degree** [ ] **DipHE**

Do normal Stage II rules apply [Y] [Y or N] - **if YES go to section** 4 **but note G below**

A [ ] Number of module credits to be passed for Honours including __ in __ and __ in __.

B [ ] Minimum number of acceptable module credits

C [ ] Maximum number of terms

D [ ] Normal number of terms of full-time study

E [ ] Maximum number of module credits allowed if study extended beyond D above.

F [ ] Honours on best [if applicable]

G [ ] Basic modules allowed, including .... which must be passed.

4 **Exemptions** XXXXXXXXXXXXXXagreed:*

	1	2	3	4	5	6	7	8
Module number:	4204	4201	4203	9201	8301			
FC's initials:								

	9	10	11	12	13	14	15	16
Module number:								
FC's initials:								

5 **Exclusions** sought/reported/agreed:*

	1	2	3	4	5	6	7	8
Module number:								
FC's initials:								

6 **Authorisation:** Senior Tutor's signature.................... Date..............

Chair of Modular Admissions Committee....[signature].... Date...17/3/88...

Amendments [if any]..

Figure 4.2 Example of form calculating a student's exemptions

OXFORD POLYTECHNIC MODULAR COURSE - Admission with Exemption   Form M10

1 Stage II Entry Date January 1988........[Month and Year]   DCM minute number: 9 2 8

Surname........................ Forenames. Nikola..

Field(s).. History & Politics ....   Date of Birth.. 25.09.1967 ....

2 **Approval for admission is** XXXXXXXXXXXXXXXXXagreed, XXXXXXXXXXXXXXXXXXXXXXXX [**ring word * required**]

....

XXX on the basis of the following **previous educational experience:**

Year(s)	Level	Subject	Institution
1986	3 'A' levels	Eng (B) German(E)	King's School, Ottery St Mary
		Economics (D)	
1987	Stage I Degree	History, Politics	University of Ulster
		American Studies	

Other relevant experience...

English Language requirement satisfied? Y [Y or N]   **CNAA** approval needed N [Y or N]

3 **Proposed/agreed Stage II programme is for** [Y] **Honours** [ ] **Degree** [ ] **DipHE**

Do normal Stage II rules apply [Y] [Y or N] - **if YES go to section 4 but note G below**

A [ ] Number of module credits to be passed for Honours including __ in __ and __ in __.

B [ ] Minimum number of acceptable module credits

C [ ] Maximum number of terms

D [ ] Normal number of terms of full-time study

E [ ] Maximum number of module credits allowed if study extended beyond D above.

F [ ] Honours on best [if applicable]

G [2] Basic modules allowed, including ..7605.................. which must be passed.

4 **Exemptions** sought/reported/agreed:*

	1	2	3	4	5	6	7	8
Module number:	2400	2404	2408	7602	7604			
FC's initials:								

	9	10	11	12	13	14	15	16
Module number:								
FC's initials:								

5 **Exclusions** sought/reported/agreed:*

	1	2	3	4	5	6	7	8
Module number:								
FC's initials:								

6 **Authorisation:** Senior Tutor's signature..........................   Date..............

Chair of Modular Admissions Committee....[signature]....   Date..11/12/87....

Amendments [if any]..

Figure 4.3 Second example of student exemptions

Stage II entry it has been agreed that she may 'trail' it in that stage, i.e. she will include it along with her advanced modules as 1 of the 2 basic modules which may also be counted.

Counselling

General principles

The overall aim of counselling is to provide students with readily available information and advice before and throughout the Course. Each student has substantial scope for individual decision-making, and it is vital that this is exercised on an informed basis, fully utilizing the qualities of choice and flexibility inherent in the Course. Students are helped to maximize their performance, to cope with academic and other difficulties that may arise and to prepare for the future beyond the Course.

The counsellors

All enrolling students are assigned to a personal tutor, normally an academic teaching in one of their fields, who has responsibility for a total of about a dozen students in various years of the Course. This relationship lasts for the duration of the Course, unless either party seeks a change. The personal tutor offers academic advice on the student's choice of programmes as well as other more general advice as required. The tutor is able to monitor each student's progress by the termly receipt of a copy of the latter's programme, and the tutor also receives copies of any letters sent to his/her students from the registry, for example after examinations committee meetings.

Senior tutors offer information and advice to applicants both on an individual basis and on collective occasions such as visit and interview days. In order to maintain a link between admissions and progress in the early part of the Course, senior tutors (along with personal tutors) are responsible for counselling all students in Stage I of the Course. They too receive termly records of their students, check on their progress, and make recommendations to the Stage I examinations committee when appropriate; letters subsequently issued from the registry are copied to them.

Field Chairs advise individual enquirers and talk to all those prospective entrants to their fields who attend a visit day. Field Chairs (along with personal tutors) are responsible for counselling all students in Stage II, and are systematically involved in comparable ways to those detailed above for senior tutors. The Field Chair's general administrative responsibility for the subject is reflected in liaison with student representatives from all years on the field committee and, in many cases, the production of a field guide offering advice on various matters.

The Modular Course Co-ordinator plays a major counselling role in relation to prospective applicants. Many make individual contact by letter, telephone or in person, among them mature applicants, those with deficient or non-standard entry qualifications, and those seeking admission with exemption. The Co-ordinator gives general explanatory talks on visit and interview days and is active in schools liaison work, responding to requests to attend events such as county higher education conferences and to visit individual schools and colleges. The Co-ordinator acts as personal tutor to the large number of associate students. He (and the Assistant Dean) also counsel many other students on the Course who consider that they have problems, and also respond to queries from other staff as to how best to advise individual students.

A range of specialist advice is available to Polytechnic students in general through Student Services. In addition to counselling on particular personal problems, advice is available on aspects such as accommodation, study skills and careers. The dissemination of information on employment is facilitated by a Careers Liaison Group which provides a forum for student services staff and Modular Course tutors. In 1987–88 a new module, Graduate Careers, was introduced, designed to offer a new way in which student consciousness of this important issue can be raised.

All staff in their counselling work can draw on a range of standard publications, most importantly the *Staff Guide* which offers detailed guidance on procedures and expected roles.

Choosing programmes and changing programmes

Choice and flexibility are two much emphasized and real features of the Modular Course. Although all things are not possible all of the time students possess a considerable measure of initiative in programme design. The system is designed to see that student decision-making is well informed.

Students are actively involved prior to enrolment. In late summer those that have indicated their wish to accept an offer are asked to make an input into their Stage I programme. While 6 or 7 modules are usually predetermined, as compulsory for their chosen field(s), they are invited to choose the remainder, up to a total of 12. They are advised to choose the compulsory modules for an extra field, not only to encourage breadth of study but also to serve as a foundation for a possible change of field in the future. That choice made, they are further asked to choose up to 12 individual modules. On the basis of this return the computer produces a Stage I programme, incorporating these choices as far as practicable, and a term 1 timetable. From enrolment onwards the student can seek to renegotiate the registered programme. For a minor change in a single term Form M99 is used (Figure 4.4); for more fundamental changes, or the construction of a complete programme, an M100 (Figure 4.5). Vital

oxford polytechnic
modular course

ADDITION or DELETION of Modules
Registration of a change to a programme

Form M99

Student's surname .... forename ....DAVID....................

fields ....AE / H1.................................... (Field change requires approval using form M101)

Student number

3	5	1	5	4	2	C

Enter all 7 digits
– see note overleaf

Ex: 1 9 8 5 – 8 6

Academic Year: 1986-87

TERM (1, 2 or 3) 3 — use a separate form for each term affected

For office use [ ] Tt

Modules to be DELETED

Withdrawal from a module is not permitted after week 6 of term

Module number (4 digits)	Termcode	MODULE TITLE	Slot number(s)
8203	3	ORGANISMS AND ENVIRONMENT	11,12,13

Modules to be ADDED

Modules may **NOT** be added after Week 1 of term (Week 2 for students in their first term of Stage I).
Use Form M100 for registering a complete term's or year's programme.

Module number (4 digits)	Termcode	MODULE TITLE	Slot number(s)
6203	3	INTRODUCTION TO TEACHING & LEARNING ACROSS THE CURRICULUM	5,6

IMPORTANT The student **MUST** check that no slot clash will arise from any added modules (consult current list of modules running available from the Modular Office) and **MUST** take his latest Student Record with him when collecting the following signatures.

Personal Tutor' signature (All programmes) .... Date 18/3/87

Senior Tutor's signature (Stage I programmes) ....RM Bainbridge.... Date 20/3/87

Field Chairmen's signatures (Stage II programmes) .... Date ....

.... Date ....

Student's signature .... Date ....

Figure 4.4 Form used to register a student's change of module (M99)

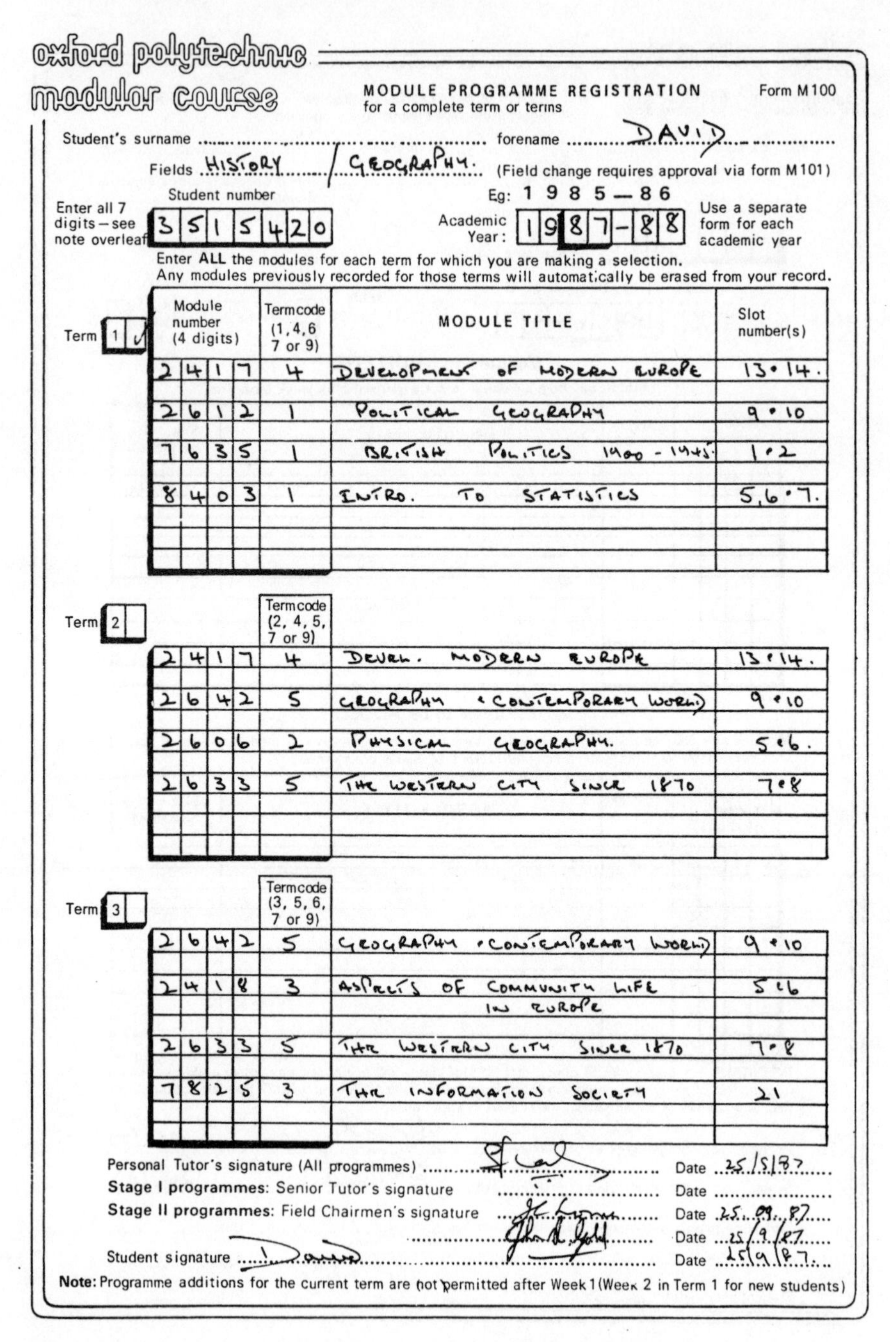

oxford polytechnic modular course

MODULE PROGRAMME REGISTRATION for a complete term or terms — Form M100

Student's surname .... forename DAVID

Fields HISTORY / GEOGRAPHY. (Field change requires approval via form M101)

Enter all 7 digits – see note overleaf — Student number: 3 5 1 5 4 2 0

Eg: 1 9 8 5 – 8 6 — Academic Year: 1 9 8 7 – 8 8 — Use a separate form for each academic year

Enter **ALL** the modules for each term for which you are making a selection.
Any modules previously recorded for those terms will automatically be erased from your record.

Term 1 ✓

Module number (4 digits)	Termcode (1, 4, 6, 7 or 9)	MODULE TITLE	Slot number(s)
2417	4	DEVELOPMENT OF MODERN EUROPE	13·14.
2612	1	POLITICAL GEOGRAPHY	9·10
7635	1	BRITISH POLITICS 1900 - 1945	1·2
8403	1	INTRO. TO STATISTICS	5,6·7.

Term 2

Module number	Termcode (2, 4, 5, 7 or 9)	MODULE TITLE	Slot number(s)
2417	4	DEVEL. MODERN EUROPE	13·14.
2642	5	GEOGRAPHY & CONTEMPORARY WORLD	9·10
2606	2	PHYSICAL GEOGRAPHY.	5·6.
2633	5	THE WESTERN CITY SINCE 1870	7·8

Term 3

Module number	Termcode (3, 5, 6, 7 or 9)	MODULE TITLE	Slot number(s)
2642	5	GEOGRAPHY & CONTEMPORARY WORLD	9·10
2418	3	ASPECTS OF COMMUNITY LIFE IN EUROPE	5·6
2633	5	THE WESTERN CITY SINCE 1870	7·8
7825	3	THE INFORMATION SOCIETY	21

Personal Tutor's signature (All programmes) .... Date 25/9/87

Stage I programmes: Senior Tutor's signature .... Date ....

Stage II programmes: Field Chairmen's signature .... Date 25.09.87

.... Date 25/9/87

Student signature .... Date 25/9/87

Note: Programme additions for the current term are not permitted after Week 1 (Week 2 in Term 1 for new students)

Figure 4.5 Form for module programme registration (M100)

oxford polytechnic modular course

AGREEMENT TO CONSIDER A CHANGE OF FIELD

Form M101

PROCEDURE FOR CHANGE

1) Discuss your intentions with your Personal tutor and then with the Field Chairmen, and when you are sure that you want to change, obtain their signatures on this form to request the change. You are advised to obtain the signatures from the Field Chairmen of your new proposed fields first.

2) Hand the form into the Modular Office by the end of Term 1 for changes requested at Christmas, end of Term 2 for changes requested at Easter, end of Term 3 for changes requested before the next Academic year.

3) The request is considered with all others by the subject Examinations Committee of your proposed fields in the light of your performance and the number of students in the fields concerned. You will be informed of the decision at the start of the term following the request.

4) If there are any special reasons for requesting an immediate change, make sure that they are given by the Field Chairman.

5) You are not normally allowed to change fields in Term 1 Year 1.

Student - Complete this area only

Surname ....

Forename(s) ..DAVID..........................

Number 3 5 1 5 4 2 0

Academic Staff to complete this area:

		Field Name	Field Code	Field Chairmen's signatures:	Date(s)
PROPOSED FIELD(S)	1	HISTORY	HI	[signature]	25.5.87.
	2	GEOGRAPHY.	GG	[signature]	25/9/87
CURRENT FIELD(S)	1	APPLIED EDUCATION	AE	[signature]	18/9/87
	2	HISTORY.	HI.	[signature]	25.9.87.

Student's signature ....[signature]....

Date ...18/9/87.......

Personal Tutor's signature ....[signature]..........

Agreement to consider in (Tick one box only)

- Christmas Vacation ☐
- Easter Vacation ☐
- Before Term 1 next year ☐
- Immediately (give reasons) ☑

Important: Field Chairmen must check whether students who are expecting to complete Stage I within one year are still able to do so. If compulsory modules cannot be completed within one year then Examinations Committee approval will be required unless entry to Stage II is delayed.
Stage I must be completed within two years.

Reason(s): ....

....

Conditions Proposed (if any) ..Dispensation from 2601; must pick up a.... 2606:2 in 1987-8 [signature]

Conditions agreed by Chairman, Examinations Committee ....

Any change of programme consequent upon this change of field must be notified separately on forms M100 available from the Modular Office. No field change is effective until **approval** is granted.

M101 revised 02-87-(500)

Figure 4.6 Agreement to consider a change of field (M101)

common features of the two forms are the signatures required on them to establish their validity for implementation. The signatures aim to ensure that all decisions are discussed with the relevant staff and their implications realized. In addition to the student's signature any change must bear the signature of the personal tutor plus, in Stage I, that of a senior tutor, or, in Stage II, those of (usually) two Field Chairs. Choice of Stage II programmes is required of Stage I students by a common deadline date, week 7 of term 2 in what is normally their first year, so that student numbers on modules, and therefore resource requirements, for the following year can be forecast in advance. Constant feedback of assessment results is one factor encouraging students to adjust their registered programmes – about 30 per cent of students make some change every term. There are certain time limits on the making of changes. Students may register for an additional module up to the end of week 1 of the relevant term (week 2 for new students in their first term), and must officially delete modules by week 6 (about to be brought forward to week 4) of the current term. The effect of the pre-enrolment registration of Stage I and of the term 2 registration of Stage II modules is that all subsequent minor changes are to an agreed, full programme. In this respect the 'cafeteria' metaphor for describing the Course is quite misleading.

Changing fields

Most students make use of the facility to change their minds about whether or not to take individual modules. The Course also goes a long way to accommodate the not uncommon student experience of making a mistaken choice of degree subject. The possibility that a student may wish to change a field of study is anticipated by the initial advice to take the modules compulsory for an extra field during Stage I. Many students do subsequently seek a permanent change to this alternative field. The reasons may be negative, in that they find themselves less attracted to, or less successful in, an original choice than they had expected, or positive in that experience of the alternative (and to some an entirely novel) field develops changed interests.

Field changes are not allowed in the first term of Stage I but may be sought on the appropriate M101 form thereafter (Figure 4.6). Again the need to secure the signatures of the personal tutor and all Field Chairs involved ensures that the implications are fully appreciated. The critical decision is that of the Chair of the field which the student wishes to enter. The Chair's decision will be based on an assessment of the student's academic suitability and on the estimated size of the Stage II group. Positive decisions can be taken from January onwards but, especially for fields for which there is much pressure to enter, decisions are commonly held over until July when the student's whole Stage I record is known. The imbalance of supply and demand in popular fields results in Chairs setting guideline conditions to be met – for example, in order to get a favourable decision the student must not merely pass the field's compulsory modules but should achieve grades of say, B, B, and B+. Exceptionally, students are permitted to enter a new field and enter Stage II despite not having taken (or passed) one or two compulsory basic modules – such 'trailed' modules must be passed in Stage II. The number of students successfully changing fields has risen in recent years and now stands at 30 per cent of the intake. For some the change means spending a second year as a part-timer in Stage I passing outstanding compulsory modules. The great majority of changes are from one single field to another, but there are rare cases of students who change both fields. Field changes become increasingly difficult once Stage II is entered, but they are permitted in exceptional circumstances, for example, when the first school experience convinces an applied education student of the need to change to the education field.

5

Assessment and Examinations

David Scurry

Principles

The Modular Course, by its very nature, subjects students to the process of progressive assessment which is one of the three main principles referred to in Chapter 1. Assessment is at least termly and, in practice, through the regular use of coursework assignments, takes place at more frequent intervals. Progressive assessment relates directly to the principle of credit accumulation in that passes in individual modules are built up progressively towards a final award. It also relates indirectly to notions of responsibility and choice in that the Course aims to allow students to make informed choices concerning their future programme of study based on feedback from past assessments.

Modules

The basic unit of credit is the module and in order to try to standardize its credit value across the many subjects and disciplines of the Modular Course it is defined in terms of notional student effort.

A basic module is 100 hours of student effort, an advanced module, intended to allow a greater depth of study, 120 hours. Most modules are of one term's duration, the Stage I student taking 4 basic modules a term and a Stage II student taking, on average, just over 3 a term, usually 10 a year. In defining modules in terms of student effort a basis is achieved for comparing arts and science modules with their differing class contact time. Modules may also be double in size and/or taught over more than one term. The single module is, therefore, a flexible unit of credit which can be adapted to the different academic demands of the variety of subject areas on the Course.

In 1988 the range of modules included two-term doubles, one-term doubles (used for school experience on the BEd), two-term singles,

Table 5.1 Number of modules on offer 1987–88 (excluding project, dissertation or independent study modules)

Size of module – 1987–88	
One-term single	480
Two-term single	19
Three-term single	12
One-term double	2
Two-term double	57
Three-term double	16

three-term singles (used in music performance modules) and three-term doubles (used in languages). The Course abandoned the use of triple and quadruple modules at the time of the 1984 CNAA resubmission for reasons that had to do both with inflexibility for student programmes and with the risk of student failure. Failure in the quadruple module, even by only 1 per cent, meant all four credits were lost. The Course decided, therefore, that from 1984 only single and double credits would be allowed, the vast majority following the one-term single pattern (see Table 5.1).

Two-term double modules are used mostly in humanities and social sciences and give longer periods of study for students over broader subject areas. However, over the period since 1986 there has been a noticeable move in these areas towards splitting double modules into two singles. Examples include law, sociology, economics, politics and English literature. There has also been a move in history away from two-term single modules, which were designed to give students a lower workload each week and more reflective time, to either standard one-term singles or two-term doubles.

At the time of the 1984 resubmission the MCC also discussed a move towards a semester system. This was seen by some to have advantages by conforming to the system adopted by most other modular and course unit schemes in polytechnics and universities. It would also give teachers more time on each module and reduce the number and frequency of examinations. The MCC voted by a large majority to retain termly modules for reasons including greater flexibility in module size and length and more frequent feedback to students on their progress. The termly pattern also allows all examination business to be conducted over the vacations prior to the start of the subsequent term and, therefore, students are able to assess their position and make informed choices and decisions on their future programme of study. Final, moderated marks and grades are given to every student before they embark on their next series of modules. The termly style of teaching is also used for non-modular courses and staffing of both types of courses is easier across the Polytechnic when all courses use the same pattern.

Programmes

In September 1988 there are 38 single and 6 double fields on the Modular Course allowing 709 different combinations of subjects for certificate, DipHE, degree and honours degrees including BA, BSc and BEd. There can, therefore, be said to be 5,672 courses available on the Modular Course. Over 800 modules are available to students and it is probable that all 3,000 students registered have unique programmes of study.

A student programme will include compulsory modules as well as some of those recommended by fields. It will also include modules chosen by the student from any other subject across the Course. All modules are open to any student possessing the stated prerequisites although there are regulations for Stage II preventing students taking basic modules for which they have an equivalent qualification. The Course has striven forcefully over the years of increased student demand and decreasing resources to preserve the notion of open access to modules in order to maintain the modular principles of real choice and flexibility.

Assessment

Assessment of students on the Modular Course takes place almost exclusively within modules. There are no overall Course rules concerned with how modules are to be assessed and a wide variety of assessment methods and styles have been developed by fields and individual module leaders.

Module descriptions must specify how a module will be assessed in terms of the proportion of examination to coursework and can only be changed via the internal validation procedure described in Chapter 2. Any variety of examination/coursework proportions are possible and the majority of modules are assessed either by 100 per cent examination or 100 per cent coursework or the following proportions of examination to coursework: 80/20, 75/25, 70/30, 60/40 or 50/50.

Many fields have a policy that groups of similar modules are assessed by the same weighting of examination to coursework. In biology, for example, modules with practical classes are usually assessed 75/25, the coursework component consisting of laboratory reports. In English literature all Stage II students are required to study linked core and options modules from at least one historical period. All the core modules run in term 1 and are assessed 100 per cent by examination. The examination requires an essay on one of the themes of the period concerned and a commentary on a piece of text. The option modules carry a double credit and run over terms 2 and 3. They cover three detailed studies and include an extensive essay comprising 40 per cent of the assessment. The examination, counting for the remaining 60 per cent, is based on questions relating texts to the overview given in the core: text in context.

The Course defines examinations as formal, invigilated, end of term assessments which can take the form of unseen or seen papers. The *Staff Guide* recommends that the length of examination be dependent on the size of module and the examination/coursework proportion. Most single modules have a two-hour examination unless the examination counts for 50 per cent or less of the overall assessment, in which case one hour is the recommended maximum. All other forms of assessment are included in the term coursework. This may include essays, seminar papers, seminar presentations, projects, case studies, laboratory work, performances or exhibitions. Coursework may be carried out individually or in groups; the latter may involve peer assessment. Also included in coursework are class tests which may take the form of mid-module timed assessments or practical tests in laboratory classes.

Comparison of workloads between modules and fields is difficult. Module leaders are responsible for informing students in writing at the beginning of a module of the assessment structure, the number of pieces of coursework required, their weightings and the deadline for their submission. Staff are also encouraged to specify the amount of time students should spend on different aspects of assessed and non-assessed work. This helps staff focus on the demands they are placing on students within the specified 100 or 120 hours of student effort required in each module. However, it is probably the students themselves who, through the experience of a wide variety of modules, can make the best comparison of demands, workloads and standards. The Course, fields and module leaders make frequent use of student feedback for comparison.

The tight specification of aims and objectives, what is to be taught and how it is to be assessed in each 11-week module can lead to novel, innovative and exciting teaching, learning and assessment methods. Staff have also been able to learn from each other through exposure to different philosophies of teaching and assessment in other disciplines. This takes place through direct contact with colleagues across the Course but also through their students, who are able to take modules from a variety of subject areas.

Visual studies, for instance, examine the majority of their modules by the assessment of a diary (25 per cent) and an end of term exhibition of work (75 per cent). In education all Stage I modules are assessed 60 per cent coursework, 40 per cent seen examination. The coursework consists of a group report on an educational issue requiring the students to use their organizational skills in completing the task. Each report is self-assessed after seminars in which all reports are presented and finally moderated, up to a maximum of one grade only, by the tutor.

Learning skills are developed within both the physical and human pathways in the geography field. Modules are assigned to levels, the higher levels using proportionally more groupwork and projects and, correspondingly, placing less reliance on lectures. The increased use of independent learning at higher levels within the pathway is reflected in the assessment pattern within the modules.

All students on a particular module are assessed in the same way. No distinction is made on the grounds of students' field of study, mode of study or status. The Course does not allow special questions to be set in an examination which are compulsory for some groups of students and not others. However, students are expected to bring to a module and to use in an examination perspectives related to their particular fields of study. A variety of coursework essays might be given, for example, and students encouraged to select topics in which their own field-specific knowledge would be useful.

Several modules are specifically designed to attract students from different areas of the Course. In such cases staff from the different areas may themselves become involved in teaching sub-groups of their own students exercises relevant to their particular area or specialism. Staff from hotel and catering management, for example, take practical classes of catering students on introductory computing modules and carry out exercises with spreadsheets and other commercial packages with applications to food costing or in use in hotel front office computer systems. Other departments do likewise. The computer studies staff provide the theory and background information to all students and set special exercises for students not catered for by their own department. Other modules provide separate seminar classes and exercises relevant to the different requirements of fields. Examples include physics and electronics problems in basic mathematics classes.

Most modules, however, are not designed to operate in this manner and rely on the students themselves using the module for different purposes. Introduction to economics, for example, is compulsory for economics and accounting & finance students and leads directly to further studies in these subjects. It is recommended for publishing students as it gives a useful background to advanced work. It is also a free choice option to all other students on the Course fulfilling the role of, say, a relevant introduction to economics for those on business and management fields such as tourism or catering management, or a liberal studies option for humanities and science students. On many modules staff will be aware that they are teaching students from different stages of the Course and with different aims and backgrounds.

Marks and grades

Each module credit carries an assessment weighting of 100 marks. Marks and associated grades are published for students for each module taken. Marks and grades are related as shown in Table 5.2. A progressive average

Table 5.2 Relating marks to grades

Grades

(i) As an indication of student progress and, in the case of pass grades, for inclusion in the final transcript the examinations committee shall award grades for modules and synoptic examinations according to the following scheme:

Grade	Description	*Percentage mark*
A	pass grades	70–100
B+	pass grades	60–69
B	pass grades	50–59
C	pass grades	40–49
S	Pass for modules with pass/fail assessment only	
R	Entitled to reassessment without retaking module	0–39
F	Not entitled to reassessment without retaking module	0–39
P	Pass at reassessment	40 (maximum awardable)
FR	Fail at reassessment (where the marks at the initial assessment and at reassessment differ the higher mark will be awarded)	0–39

(ii) A student awarded an R grade shall be entitled to reassessment at the end of the following term or in the vacation prior to the next academic session. Reassessment implies either re-examination or the submission of further coursework for evaluation or a combination of the two.

(iii) No more than one reassessment shall be allowed in any term or in the vacation prior to the academic session, and a student awarded R grades for two or more modules may normally choose in which module to be reassessed.

(iv) At a reassessment only grades P (40 per cent) or FR shall be awarded.

of marks gained on modules passed is printed on a student's records each term. Stage I and Stage II averages are printed separately as it is the Stage II average only which forms the basis for determining the class of honours. S grades and modules for which exemption has been granted are not included in the average. Similarly credits earned on exchange or by credit transfer (normally a maximum of four exchange credits are allowed) are counted towards the number of credits required for an award but the marks are not used for the purposes of classification. Only marks gained on the Course are used in the calculation of honours.

Resits, as opposed to fail grades, may be awarded to students, allowing them to be reassessed without retaking the complete module before the next meeting of the examinations committee. Resits may take the form of an examination, or further coursework or both, the decision usually lying with the module leader. In general students in Stage I are given resits when they do not reach a pass grade. In Stage II, however, resits may only be

Table 5.3 Statistical analysis of module results

Per cent of students deferred with grades:								Number of students with grades:												
A	B+	B	C	P/S	MC	F/R FR	DF FO RO	A	B+	B	C	P/S	MC	F/R FR	DF FO RO	No. entered	Average	Standard deviation	Module number	Title
Sociology																				
0	36	43	1	0	0	0	19	0	24	29	1	0	0	0	13	67	58	5	78043	Sociological analysis
0	20	40	0	0	0	20	20	0	1	2	0	0	0	1	1	5	49	20	78233	Sociological theory II
0	42	24	16	0	3	5	11	0	16	9	6	0	1	2	4	38	55	10	78253	The information society?
18	36	9	18	0	0	9	9	2	4	1	2	0	0	1	1	11	55	19	78283	Industrialization and development
17	17	33	33	0	0	0	0	1	1	2	2	0	0	0	0	6	58	12	78515	Sociology of work
0	60	20	0	0	0	0	20	0	3	1	0	0	0	0	1	5	63	5	78889	Independent study in sociology
10	60	20	0	0	0	0	10	1	6	2	0	0	0	0	1	10	62	6	78909	Sociology interdisciplinary dissertation
33	33	33	0	0	0	0	0	1	1	1	0	0	0	0	0	3	65	8	78999	Sociology dissertation
								Subject group summary – all modules:								145	57	9		
								Basic modules:								67	58	5		
Law																				
2	21	27	21	2	0	22	5	3	26	34	26	3	0	27	6	125	46	18	79055	Introduction to law
0	11	27	37	3	0	10	13	0	7	17	23	2	0	6	8	63	48	9	79063	Constitutional law
3	19	42	17	0	3	11	6	1	7	15	6	0	1	4	2	36	53	8	79123	Crime and society
6	42	29	15	0	0	2	6	3	20	14	7	0	0	1	3	48	57	8	79233	Civil obligations
0	17	22	42	0	3	8	8	0	6	8	15	0	1	3	3	36	49	8	79243	Civil liberties

0	40	40	7	0	0	13	0	0	6	6	1	0	0	2	0	15	56	10	79909	Law interdisciplinary dissertation
29	14	43	14	0	0	0	0	2	1	3	1	0	0	0	0	7	59	10	79999	Law dissertation
Subject group summary – all modules:																374	51	13		
Basic modules:																188	46	16		
Biology																				
5	12	23	26	14	0	14	7	2	5	10	11	6	0	6	3	43	48	10	81043	Fundamentals of human physiology
3	26	27	27	1	0	13	3	2	19	20	20	1	0	10	2	74	50	13	81133	Molecular biology of the gene
9	23	23	25	0	0	16	5	4	10	10	11	0	0	7	2	44	51	15	81163	Separation and analytical methods in biology
14	16	37	22	0	0	10	0	7	8	18	11	0	0	5	0	49	54	13	81223	Circulatory, respiratory and renal physiology
11	28	17	22	0	0	22	0	2	5	3	4	0	0	4	0	18	50	15	81243	Sensory physiology
15	35	31	16	0	0	4	0	8	19	17	9	0	0	2	0	55	57	11	81273	The health of man
14	64	21	0	0	0	0	0	2	9	3	0	0	0	0	0	14	64	5	81293	Environmental and work physiology
8	25	58	8	0	0	0	0	1	3	7	1	0	0	0	0	12	57	8	81443	Microbial physiology
25	18	25	18	0	3	5	8	10	7	10	7	0	1	2	3	40	60	15	81503	Immunology
26	48	11	7	0	0	4	4	7	13	3	2	0	0	1	1	27	63	9	81909	Interdisciplinary project in biology
0	0	0	100	0	0	0	0	0	0	0	1	0	0	0	0	1	*	*	81989	Project in biology
12	47	29	6	0	0	0	6	2	8	5	1	0	0	0	1	17	63	12	81999	Project work in human biology
Subject group summary – all modules:																394	54	13		
Basic modules:																40	40	16		

given on compulsory modules, normally including projects and dissertations. There is no requirement to give resits but staff are advised to treat Stage I students generously. Nevertheless, absence from an examination or the non-completion of coursework often results in the award of a fail grade.

Students may take only one resit at the examination session immediately following the one in which the resit was awarded. The rule allowing students only one resit per session has been developed for both academic and logistic reasons. It is felt that allowing students more than one resit is too generous, unnecessary when they can often make up lost ground in subsequent terms, and allows the examinations office to timetable all resits in a single block at the beginning of examination week.

Resits may not be carried forward to another examination session, even if the particular module is repeated and another resit examination offered. Students awarded more than one resit must choose, in consultation with their personal tutor, which one is more important to take. They may, of course, choose instead to retake the module if it is offered again in a subsequent term. A pass at resit for all modules carries a mark of 40 per cent only. However, students taking resit examinations for exceptional reasons, such as an examination missed through illness, are usually allowed to count the full marks earned.

The Course has debated the use of a grades-only system of marking to replace marks as the basis of assessment. This was rejected on grounds of ease of use and familiarity with the 100 per cent marking scale. Nevertheless, it is acknowledged that disciplines maintain different marking conventions – differences that a grades-only system could reduce. Well known examples include the rare use of the 80–100 per cent range in arts and social studies and the full use of very high and very low marks in science and mathematics. Despite these traditions, analysis of student results does not indicate that it is more difficult to gain a first in arts than it is in science subjects. However, students gaining good degrees in the arts tend to have a very flat profile of marks clustering around the 70–75 per cent range, whereas science students show more peaks and troughs with marks often ranging from 65 to 85 per cent.

A full statistical analysis of module results is produced each term (see Table 5.3). For most modules the mean mark will normally be in the low to middle 50s and if there are more than 30 students the standard deviation will tend to be between 10 and 20. Modules with very high or low means or wide or narrow standard deviations are referred to field committees for comment in the subsequent annual review cycle.

Examination week and Examination Committee

Formal invigilated examinations take place in week 11 and a full examinations board (the MEC) is held after the end of each term. The Course is committed to producing complete, moderated assessments at the end of each module in order that students have full information on their progress before commencing their next set of modules. Under this system students are also able to graduate as soon as they have met the requirements of an award and do not have to wait until July each year.

The examination timetable itself is based on the slotting system and clashes are, therefore, minimized. The Course uses 12 different paired timetable slots for most modules (see Appendix I) and it is, in theory, possible to examine all modules over six working days from Saturday of week 10 through to Friday of week 11 with two examination sessions a day. However, logistic problems and slotting practice have made this ideal difficult to achieve. Due to the growth of the course, there is not now examination space in the Polytechnic to examine the total number of students registered for modules in each examination session. Many fields now repeat their more popular modules in separate timetable slots and these students cannot then be examined within the normal week 11 period. These developments have led to the encroachment of examinations into Thursday and Friday of week 10 but, nevertheless, over 7,600 examination places were administered by the examinations office over an eight-day period in July 1988.

Double marking raises problems for staff already under great pressure of examining three times a year and also working within the tight deadlines of examination week. Some feel that there is no need to double mark all modules as a large number of examinations contribute to the student's final award. The students are already assessed independently by a larger number of staff than under traditional degree systems. Others feel that double marking is essential in order to be fair to students and external examiners. The *Staff Guide* offers advice on this matter and recommends to staff when a module should need to be double marked.

For a module involving several different forms of assessment and/or having more than one member of staff teaching and assessing it, double marking is not normally required. Double marking is appropriate for a module having a high proportion of assessment in one form (for example, 100 per cent examination, dissertations and projects) or if assessment is not objective or quantitative (such as a performance or presentation). Sample double marking by a second internal examiner is normally sufficient for modules having more than 50 students.

As soon as scripts are marked by internal examiners they are sent to external examiners and the unmoderated results fed into the student management system. The deadline for submission of mark sheets is the Tuesday of week 12. This short deadline is achieved through the dedication of staff and the strategy of placing modules with large student numbers early in examination week. This allows staff with the largest workload most time. Student records are run off overnight and by Wednesday morning a full list of results by student in field order is available for collection by Field Chairs for Stage II students and by senior tutors for Stage I.

This procedure of distributing results to approximately 55 staff across the Course enables the progress of the 3,000 students to be reviewed very quickly. Recommendations on student progress are made to the central management team of the Dean, Assistant Dean and Co-ordinator by Field Chairs and senior tutors before noon on Thursday. In practice no action is taken for those making satisfactory progress, but recommendations are made, in the form of a series of standard letters, where special action is required in respect of individual students. The standard letters cover typical student problems such as failure to complete various stages of the Course and warnings. On the more positive side, degree classifications for graduates are also recorded, based on the students' progressive average.

Recommendations returned by each Field Chair and senior tutor are then collated onto a master set of all student records by the management team. On Friday of week 12 the examinations office extract all records of students with such recommendations, divides them into three groups (Stage I, continuing Stage II and graduating students), and these records are printed as the examination booklet which is available for collection by staff centrally involved in examinations committee business by noon on Monday of week 13.

From the 3,000 students on the Course typically 300 records with recommendations will be presented to the examinations committee via this booklet. In addition 20–50 records of graduating students will be listed at Christmas and Easter and approximately 700 in July.

A typical Christmas examination timetable is shown in Figure 5.1. An Easter examination schedule would follow the same pattern while in July the examinations committees would meet on the Tuesday and Wednesday of week 13 rather than immediately prior to the beginning of the following term.

The Modular Course uses a two-tier examinations committee structure, the overall strategy being to provide both tiers with relevant information and recommendations in such a way that examiners can concentrate on the important issues relating to student progress. This includes students who are moderated from pass to fail as well as those on the borderline between two classes of degree.

The first tier to meet is the subject examinations committees. All subject committees meet on the same day. The Chair of each committee is usually the Field Chair or Head of Department and membership consists of module leaders from the field and the external examiner. The subject examinations committees are supplied with records of all their students and the examinations booklet. Additional information supplied by the Modular Office includes details of medical certificates received over the term, field change requests from students and statistical summaries of module results.

The subject examinations committee discusses moderations by the external examiner which may, of course, affect recommendations published in the examination booklet. Individual modules are discussed, as is

```
OXFORD POLYTECHNIC                                        MCC 87/49
MODULAR COURSE - EXAMINATION ARRANGEMENTS FOR TERM 1 1987/88
=============================================================

- DECEMBER ----------------------- WEEK 10 ---------------------------------
  ====
Friday 4      4.30 pm DEADLINE for the submission of marks by internal examiners
              to the Examinations Office for modules examined up to the end of
              week 9 (ie.  including ALL project/dissertation modules)

---------------------------------- WEEK 11 ---------------------------------

              Main Examination Week

---------------------------------- WEEK 12 ---------------------------------

Tuesday 15    NOON DEADLINE for the submission of ALL remaining marks to the
              Examinations Office.

Wednesday 16  Late AM - FIELD CHAIRMEN and SENIOR TUTORS - collect your student
              record printouts together with lists of names (form M9 or M11) onto
              which you must report your recommendations on problem students.
              N.B. NIL returns are required.

Thursday 17   10.00 AM - DGAS/JEB to mark-up and create the Graduating booklet.

              2.00 PM DEADLINE for FIELD CHAIRMEN to return to the Modular
              Office forms M9 (NIL Returns required) with their recommendations
              for action on continuing students.

              2.00 PM DGAS to mark-up and create the continuing booklet.

              2.00 PM DEADLINE for SENIOR TUTORS to return to the Modular Office
              their recommendations on Stage I problem students, using forms M11.

 Friday 18    3.00 PM - STAGE I MARK-UP in CG 09.

---------------------------------- WEEK 13 ---------------------------------

Monday 21     2.00 PM  ALL booklets available for collection from the
              Examinations Office. (The booklets may be ready beforehand. Please
              check on availability with the Examinations Office on Ext. 3032).
          NB  Chairmen of Subject Examinations Meetings should also collect forms M7.

- JANUARY 1988 ------------------ WEEK 0 ---------------------------------

Monday 4      12.30 PM - Lunch for members of MCC and external examiners.
              5.00 pm DEADLINE for the return of forms M7 to the Modular Office
              and Moderated mark sheets to the Examinations Office following
              Subject Examinations meetings.
              4.30 to 5.30 Chief External Examiners available for consultation on problem
              cases in C105 (Chairman's Office).

Tuesday 5     2.00 pm   Modular Examinations Committee meeting in G1 16.
```

Figure 5.1 A typical Christmas examination timetable

the overall progress of students and degree classifications. Any decision to alter, delete or add a recommendation in the light of moderations is reported to the Assistant Dean, as Chair of the MEC by 5.00 pm.

This information is fed into the second tier of the process, the MEC. The Assistant Dean, as Chair of this Committee, uses the information from each subject examinations committee to amend his version of the examination booklet and to obtain records for any additional students with recommendations.

The MEC meets the next day and comprises the chair of each subject examinations committee and three chief external examiners, chosen to serve for two years from among the external examiners, although all external examiners have the right to attend. The chief external examiners, with the Assistant Dean, also make themselves available the previous day to discuss, with subject examinations committee chairs, students in difficult situations. This avoids lengthy discussions in the large MEC.

The MEC discusses only those students for whom a changed recommendation has been made. This includes moderations from the external examiners, late notification of medical problems, decisions on field changes which affect a student's progress or the recommendation of a higher class of degree than that indicated by a progressive average. In the latter case judgements are always made on the student's overall profile of achievements and no attempt is made to alter marks on modules to produce artificially a higher average. All recommendations in the examination booklet which are not discussed and not altered are taken as agreed. The MEC usually meets for 1½–2 hrs and may discuss or note changes to around 50 student cases.

In the week following the July meeting of the examinations committees the Modular Office prints new student records or transcripts for graduates and sends them to home addresses. At Christmas and Easter records are printed over the next three working days ready for collection by students on the first day of term along with the two to three hundred letters from the Assistant Dean.

External examiners

External examiners work very hard for the Modular Course and play an important part in the progressive assessment of students. Each is responsible for a specific set of modules and associated field. They meet field teams in the subject examinations committees three times a year to moderate examinations and to review the progress of students. External examiners are entitled to attend and contribute to the MEC. In addition the external examiners review and comment on examination papers, moderate coursework where this constitutes more than 50 per cent of the assessment of a module, review other coursework if necessary, viva students if required, recommend awards, write annual reports and advise staff on the delivery of the course.

Progression of students

The role of the MEC is to consider students' progress at the end of each term. Formal checks and reviews of student records take place at frequent intervals and, often, problems can be identified in time for remedial action and counselling.

Students progressing satisfactorily in the standard full-time mode of study, holding a local education authority grant, will graduate at the normal time. Students studying part-time, and by definition receiving no grant, will graduate on completion of the programme of study planned and agreed between the student and his or her personal tutor. Such students graduate at the point when they meet the regulations for the specific award, which is not necessarily July in any particular year. On graduation all students receive a transcript detailing the award, each module passed and the grade earned. This applies equally to certificates, diplomas, degrees or honours awards.

Students making poor progress exercise the examinations committees for a disproportionate amount of time in comparison with their number. However, in accordance with the principles of credit accumulation, full-time students can make use of Course regulations to take additional modules, at their own expense in the part-time mode, to get back on track. If successful, such students will be recommended for reinstatement of their grants and return to full-time study. Students are also able to extend their course to take extra credits should they not meet degree requirements within the standard three or four years, again by returning part-time. There is, of course, a minimum number of modules that must be passed each year for any student to be allowed to continue on the Course. This minimum is 3.

Under Course regulations all students have two years to meet the Stage I requirements of passing 10 modules including all compulsories. Full-time students meeting these requirements at the end of year one pass automatically into Stage II. Part-time students will pass automatically into Stage II as soon as they meet the requirements, usually at the end of their second year. Those passing 10 or more modules but who fail one which is compulsory may, at the discretion of the MEC, 'trail' the module into Stage II. Such modules must be passed but count as any other basic module that can be taken in Stage II. No student can, however, be in Stage I and Stage II at the same time and no student is allowed to proceed into Stage II unless he or she has fully satisfied Stage I requirements or with the express approval of the Examinations Committee.

Full-time students who pass only 3 to 9 modules in their first year have, by right, the option to return for a second year in Stage I to take *additional* modules to meet the requirements. They are not required to repeat the full year. Their grant would normally be suspended until they meet the requirements and are able to enter Stage II. This point is often reached the following September. However, students who fail by only 1 or 2 modules are often able to complete Stage I by Christmas or Easter. If they are able to construct an approved programme of study over six terms from that date they are then eligible to enter Stage II in term 2 or 3 of their second year. Their grants would be reinstated from that date and they will obviously

oxford polytechnic modular course **Form M39**

INTERMEDIATE AWARD / FINAL AWARD / DEGREE TYPE / FIELD ORDER

Use this form if you are one or more of the following:

- a BEd/BA/BSc student not wishing to take Honours
- a BA/BSc Honours Stage II student taking 2 Neutral fields or 1 Arts and 1 Science field
- intending to receive an intermediate award of either the CNAA Certificate or DipHE
- intending to leave the course after completing a DipHE or Certificate programme
- changing the order in which your fields appear on computer printouts

Name .... Field code(s): [| |] Number [| | | | | |]

Intermediate award

L [ ] **Certificate** D [ ] **DipHE**

If you wish to be awarded the **Certificate** or the **DipHE** as an **intermediate** qualification tick appropriate box above and enter the **year** and **term** you expect to be eligible for the award:

[ ] **Year** (eg 1986-87 as "86") [ ] **Term** (1, 2 or 3)

Final award (Applies to those who are **NOT** candidates for the award of a degree with Honours)

L [ ] **Certificate** D [ ] **DipHE**

B [ ] **BA Degree only** C [ ] **BSc Degree only** A [ ] **BEd Degree only**

Degree type (Applies to candidates taking either 2 neutral fields or 1 Science & 1 Arts field)

F [ ] **BA degree/with honours** G [ ] **BSc degree/with honours**

Field Order

Z [ ] **Change the order** in which your fields appear on your student record, transcript etc.

Signed.. **Date**.................... (see over for NOTES)

Return completed forms to the Modular Office as early as possible and no later then the end of the term before the award (otherwise the conferment of your award may be delayed).

For office use only (MEC decision)

H [ ] BEd Degree J [ ] BSc Degree Processed (MEC screen) [ ]

I [ ] BA Degree K [ ] DipHE only Copy to EXAMS office [ ]

Date........................ processed

09/86/sm. → **FILE ON STUDENT FILE**

Figure 5.2 Form M39 – Student selection of intermediate awards, etc.

graduate at the end of the appropriate term. Local education authorities query this pattern of study less often than they used to as principles of credit accumulation percolate through to town halls.

In Stage II the rules are operated in a similar way. Full-time students are required to pass at least 7 modules in the first year of Stage II, their second year, to satisfy the requirements of a full-time year of study. Those passing fewer than 7 modules would have their grant suspended until they had passed the additional modules. In such cases students are required to meet half the minimum requirements for their final award before the Polytechnic recommends reinstatement of grant. Students meet this condition by passing 9 modules if they are still eligible for honours and 8 for non-honours. The remaining three terms' grant can be reinstated at whichever point the student reaches the target resulting, again, in some students graduating at Christmas or Easter after their originally expected date of completion.

Full-time students not meeting the degree requirements at the end of three years may be eligible to return to take additional modules within the maximum period of five years from entry to Stage II. However, in order to prevent students gaining an honours degree by 'attrition', a maximum number of modules are allowed to be taken, if a student extends Stage II beyond six terms, in order to retain honours eligibility. This maximum is 21 for BSc and BA degrees and implies no more than 3 failures in a programme requiring 18 passes for honours. The Modular Examinations Committee uses the '21 rule' in considering the eligibility of students to continue for honours when they arrive at the end of the third year with less than 18 passes. Those with 16 or 17 passes but meeting all other requirements, and who could not keep within the '21 rule' if they were to take additional modules are awarded a degree without honours. All other students not meeting the requirements for an award are informed whether or not they are eligible for honours should they choose to continue. These procedures work pro rata for BEd students on a four-year degree.

The MEC assumes that all students will continue for honours, if eligible, unless informed otherwise by the student. Where they have a choice students use Form M39 to make this and other decisions concerning their award (Figure 5.2).

Students may use this form to indicate whether they wish to be awarded a certificate or DipHE as a terminal or intermediate award and may even indicate the order in which their two fields are to be listed on their transcript and CNAA parchment. All fields on the Modular Course are designated arts, sciences or neutral. Students use Form M39 to indicate whether they wish to be awarded a BA or BSc should there be a choice because they are taking two neutral fields or a combination of arts and science fields.

Glasnost

The rapid feedback of marks and grades on a termly basis creates an open system of assessment easily understood by students. They are able to monitor their progress closely and to relate effort and performance to accepted standards. They are also encouraged to discuss disappointing results with module leaders as part of the learning process and, with luck, use this information to improve performance on subsequent modules.

Any one student may take modules from a number of subject areas and will make comparisons of workloads and standards. If necessary, complaints and criticisms can be made through field committees and the termly staff/student meeting to the MCC. The full statistical analysis of module results (see Table 5.3) is a public document and is scrutinized by the MCC, the MEC, and FRG, for anomalies every year.

Quick and easy checks can be made on scripts or marks sheets if an administrative error is suspected by a student when his or her marks do not live up to expectations. A formal appeal against marks can be investigated immediately results are published and, unless a module from the student's final term is involved, can be resolved before the student is considered for an award by the MEC.

Although they may never fully understand the practice and culture of the examinations committees, students on the Modular Course do not feel assessment, examinations and awards to be one of the great mysteries of life.

6

Review and Evaluation

Roger Lindsay

Some key terms

- *Monitoring*
  Maintaining course. The relatively continuous use of standard system outputs to determine the necessity for corrective action.
- *Review*
  Taking a fresh look. Periodic and retrospective reconsideration of system performance ranging over standard outputs, purpose-specific reanalysis of past performance, and new evidence.
- *Evaluation*
  Weighing the evidence. Often used as a generic term for all activities yielding evidence relevant to course quality. A preferable interpretation is: a search for evidence justifying a choice between qualitatively distinct alternatives. The most common choice in higher education evaluation is between 'change' and 'no change'. If change is necessary, evaluation should also assist in discriminating among rival possibilities.
- *Validation*
  Establishing adequacy. The processes by which a new course, or a proposal for course change, is demonstrated to be acceptable.

If validation is the establishment that course proposals are of acceptable quality, monitoring, review and evaluation are the processes which attempt to maintain and enhance the quality of courses once they are running. Oxford Polytechnic's interpretation of modular structure is analogous to a federal state, with a great deal of diversity at the subject-based periphery, and a great deal of uniformity at the administration-based centre. This arrangement offers unusual opportunities for the quality control of courses. It also poses unique problems. The opportunities mainly stem from central record-keeping which creates very large and readily accessible databases incorporating student programme choices, assessment data and resource distribution data. The problems arise from the need to apply quality control procedures to each of the various levels in the system, to interactions between levels and to inputs from central services and institutional context. It should be said that many of the problems which make quality control of elements of the course above subject level a challenge for modular courses are not a challenge for conventional courses only because insufficient comparability exists to make quality control a serious possibility.

What is required ideally is an intelligent decision-making subsystem capable of sampling feedback and effecting or signalling the need for change at every level of the Modular Course which is capable of seriously affecting quality. The subsystems shown in Table 6.1 actually exist.

Monitoring

All of the opportunities which exist for course monitoring in a monolithic, longitudinal course structure also exist at field level in the Modular Course. Difficulties are created by the following features of the Modular Course:

- acceptability of modules from other fields;
- different subsets of students on each module;
- two fields contributing to student progress (except for double field students);
- interdisciplinary modules, e.g. interdisciplinary projects;
- ease of student transfer between modules and fields;
- variation in module status (e.g. compulsory or optional);
- variation in teaching staff on modules; and
- differential overlap between modules.

It is probably obvious that each of these difficulties has an educational flip side. This is no accident, as otherwise there would be no reason to allow the difficulty to exist. Additionally, other features of the Modular Course allow most of these problems to be solved. For example, centralized record-keeping enables module enrolment and cohort details to be accurately maintained. Existence of many modules which are similar in duration and assessment pattern enables ready comparison within and between fields. Termly and yearly dissemination of assessment results, which include grade frequencies and measures of location and variability, fosters understanding of properties of mark distribution and a (sometimes competitive) interest in the behaviour of academic neighbours. In a conventional course, the quality of provision is largely confounded with the success of a particular body of students. Because the staff and student mix on modules varies so freely, quality judgements on the Modular Course have come to relate much more closely to statistical properties of modules in relation to other modules. See Table 5.3 on pp. 80–1 for a summary of the assessment information which is made available to all Modular Course staff.

Many of the features of the Modular Course which present difficulties

Table 6.1 Decision-making subsystems

R*	Evaluation of each module run by module leaders
M*	Opportunity for students to identify problems at termly field committee meetings
R*	Annual evaluation of field by field students
R	Fortnightly meetings of Modular Management and Review Committee
E	Termly Modular Course development meeting (all Field Chairs, Deans, HODs, and senior tutors invited)
M	Termly general meeting for Modular Course students (teaching staff are excluded from one of these per year)
E	Termly feedback seminar having current awareness function
M	Termly Modular Examinations Committee meeting to examine effects of assessment on student progress (held in conjunction with external examiners meeting)
R	Annual review of fields by field committees
R	Annual review of fields by departments
R	Annual review of departments by faculty and Modular Course
R	Annual review of Modular Course by chairman
R	Quinquennial review of fields and course by internal and external consultants
E	Use of Educational Methods Unit on consultancy basis
E	Commissioned studies on identified issues by (past) Evaluation Officer or (future) internal or external evaluation consultants

Primary function of each quality control mechanism identified by M=monitoring; R=review; or E=evaluation.
* indicates that an evaluation element is expected, and that its absence will incur criticism, but is not imposed.

for monitoring are design features of the Course. In many cases structures have been established which allow monitoring to straddle the subject-related fault lines which exist at field level. For example, field committees which meet at least termly, and on which students are usually well represented, include *ex officio* module leaders of all modules acceptable to, not just taught by, a field. A Stage I examinations meeting monitors the academic progress of all Stage I students; this committee largely consists of senior tutors who are responsible for groups of fields. Their duties include admitting students to the Course and providing programme advice to stage I students. Their remit thus permits, and indeed requires, monitoring of students across fields.

The MEC, on which all field committees are represented by their Chairs, exists to monitor term-by-term progress of all students. It should be noted that the possibility of trans-field comparison, which certainly operates to reduce subject insularity and promote transfer of good practice, also tends to impose a pressure towards uniformity. Such uniformity is naturally favoured by the central administrative structures, as administration of a system is easier the more similar its component elements; but pressure towards uniformity constantly threatens to erode the distinctive educational requirements of individual fields. The effects of this pressure have emerged in at least one Modular Course evaluation study (see Table 6.2, Modular Course Management Structures), and a determined attempt has been made to configure the management system in such a way that Field Chairs participate in all Course development decisions, so that special needs are recognized and special difficulties cannot be easily ignored.

It is probably obvious that comparability of unitization and assessment pattern, the frequency with which modules run (often termly for popular Stage I modules), common requirements for students seeking the same qualification (e.g. all honours students are normally required to complete a project), and central provision of course-wide performance data, all operate to facilitate monitoring and to reduce the burden it places upon staff.

Review

As with monitoring, possibilities for periodic review at field level do not differ greatly from those presented by a traditional course. Difficulties are posed for the review process by Oxford Polytechnic's version of modularity, but most of these are not intrinsically produced by modular design: for example, the requirement that students study two subjects inevitably complicates review, but is not an essential feature of modular courses and frequently exists in non-modular systems. It may nevertheless be helpful to identify some challenges to effective review:

- orthogonality of modular to departmental/faculty structure;
- size and diversity of the course;
- multiplicity of levels and services;
- use of externally derived authority to resist review;
- development of burdensome bureaucracy;
- penetration beyond documentation facade;
- collusive neglect of inconvenient issues;
- asymmetrical distribution of responsibility and power;
- bias towards assessment of adequacy rather than optimality of provision; and
- lack of interest and comparative experience among students.

Management of the Modular Course by field committees at subject level, and the MCC and its executive subcommittee, the MMRC is widely seen by staff as open, participative and democratic. In a study of 'need for institutional change' (see Table 6.2), these management structures of the

Modular Course were seen as less in need of change than any other major committees in the Polytechnic. However, the Modular Course exists within the wider Polytechnic environment which has a traditional rigid, top-down, management chain of command.

Within the Modular Course identification of areas of concern has been followed by one-off evaluation studies, or incorporation into a review cycle. Almost every major evaluation study has led to minor adjustment or substantial change. This effectiveness-centred approach has not always been accepted outside the Modular Course. For example, Heads of Department have resisted review by their own staff. Central services have not always disseminated the results of their own reviews. Deans and the directorate remained almost outside the review process until in March 1987, a decision of Academic Board determined that Heads of Department, Deans and members of the directorate should have the 'opportunity for an annual staff development discussion'.

It remains true that only teaching staff are expected to be routinely exposed to both bottom-up evaluation and top-down development. This is not to imply that areas outside review are inadequate: the 'need for institutional change' review, which was a Modular Course initiative, revealed no perceptions of urgent need for change in any area of Polytechnic management. But the success of review in some areas can lead to a keener awareness of its absence in others; and however sophisticated the review process, it can affect only those dimensions of course management to which it is applied. The medium- to long-term outlook is optimistic; the success of the Modular Course educationally, academically, and as a management system which has maintained high morale among staff during difficult times, makes it a hard example to ignore.

The most serious difficulties for 'orthodox' review are created by its inflexibility. Its frequency is usually fixed, though problems may arise more or less often than review cycles. Its format is usually standard, though much of the information collected and reported at a particular time may not be relevant to current development issues. It is organizationally unidirectional, filtering fitfully up the management hierarchy, irrespective of the extent to which it is attuned to the current preoccupations of senior management. These features can all too easily cause the annual review to be perceived as pointless drudgery, cause it to be prepared in as perfunctory a manner as possible and used as a device to obscure rather than reveal problems.

The Modular Course has begun to develop away from this stultifying orthodoxy in a number of ways:

- Top-down steering. Systems higher up the management tree identify issues upon which review is to focus. This cuts into redundancy and guarantees relevance.
- Process rather than outcome orientation. Movement in this direction was first fostered by CNAA in guiding institutions towards autonomy.

It is now thoroughly internalized within the Modular Course review process. In practice its major consequences are: (a) the assignment of responsibility for assisting a team preparing a review to individual members of committees higher up the tree; and (b) collapsing levels, e.g. by having MMRC members attend a department's review of its own fields, so that there is no need for a separate event and the process of review can be observed.

- Close coupling of review and change. Proposed change is whenever possible shown to arise from identified deficiencies. Subsequent review, where possible, tracks the consequences of preceding change.
- Minimal Form Annual Reviews. Inevitably review plays a great part in the process of quality control by external agencies. Their background knowledge of an institution is typically and corporately small, and sojourn within its walls brief. Internal reviews may be much more schematic when institutional context and course structure and assumptions are shared, as occurs on modular courses. On this model, annual reviews of fields continue to be produced, but are as spare as possible. A set of mandatory elements are agreed by ASC much of the material consisting of collated and briefly interpreted statistics, available from the central administration. Other elements are optional, or derive from the need to report on recent changes, or to justify changes which are proposed. More searching and comprehensive reviews occur quinquennially.
- Variable periodicity. Review is related more closely than in the orthodox model, to the state of the system, particularly to whether change is proposed or rapid. This variation upon standard review practice is not suitable for field reviews, but may be appropriate for example for research activities within a department.

Many other obstacles to review, such as collusive disregard of inconvenient problems, when management of an institution prefers not to recognize a problem, and a course team does not want to be seen either as having problems, or as inconveniencing management, must be addressed by incorporating external expertise into the review process at judicious points. Part of Oxford Polytechnic's response to accreditation has been to increase the involvement of external reviewers in the quality control process. This strategy is not the whole solution; for example, external reviewers may be reluctant to identify under-resourcing as a serious quality problem, when they know that additional resources are not available, just as juries would not convict for lamb-stealing in eighteenth-century England when they knew conviction would lead to death. The only conclusion that can be drawn, regrettably, is that under-resourcing is bad for quality control as well as being bad for quality.

The main input to the review process is documentary evidence. Obviously this is highly selective, and has an uncertain relationship to non-documentary reality; this might be called the problem of 'rose-tinted documentation'. The most obvious solution to the problem is student representation on review panels, or access to students by a review panel. The Modular Course employs both devices. In practice, however, students are not enthused by participative bureaucracy so that volunteers are difficult to find and are probably not representative. Few students study precisely the same pairs of subjects. And in general, student opinion needs to be interpreted with caution. Students are prone to explain undesirable features of a system in terms of its most salient characteristics. It is extremely common to find Modular Course students attributing such problems as lack of time for background reading, or difficulty in combining final year commitments and applying for jobs, to the Modular Course. This is not unreasonable, given that students have not as a rule experienced any other regime, but it can be misleading.

A healthy review system is clearly necessary and desirable. The experience of the Modular Course suggests that what must be avoided is 'Exocet thinking': review systems do not stay healthy in fire-and-forget mode – they degenerate into a perfunctory ritual. The review process is not sufficient in itself as a quality control mechanism. Even when external advice and student opinion is freely available, review practice requires supplementation. In part the justification for this claim has already been provided; but there is a further potent reason.

Review is biased towards what actually exists. This is true at many levels; for example, available documentation tends to become the subject of discussion. Similarly, the adequacy of the course as it exists is usually the implicit object of review. This neglects the fact that there may be more efficient and effective ways of achieving system objectives which remain unconsidered. The conclusion must be that monitoring and review should proceed alongside other quality control exercises which are more pragmatic, which introduce a wider range of data, and which are capable of addressing the question of optimality as well as the question of adequacy. Oxford Polytechnic's Modular Course has long accepted this conclusion. A consequence of breaking new ground is that received wisdom becomes a poor guide. Instead course development has been informed by a continuing series of evaluation studies which have sought to assist in establishing priorities and in discriminating between alternative developmental paths. The style and method of this form of evaluation will now be described. As problem-centred evaluation is rather less common than methods so far discussed, an attempt will be made to locate discussion within a theoretical framework, and to illustrate it with examples of studies which have actually been carried out.

The role and purpose of evaluation in higher education

In general, evaluation is the collection of evidence relevant to the distribution of rewards and penalties or the making of qualitative choices.

In principle the choices may be between people, objects, events, processes or structures. Discussion in this chapter will confine itself to evaluation within an established institution. In this context the object of evaluation is always a system which is usually composed of a structure and a process element. The existence of the process element means that evaluation is usually oriented towards detection of need for change rather than towards choice between temporally co-existing alternatives.

While change without evaluation is dangerous evaluation without at least the possibility of change is futile. The truth of the first of these dicta is easy to see: any change may make matters worse rather than better; only in a system which is as bad as it can possibly be, can change be introduced which must be beneficial. The second dictum may be harder to accept. Would we not evaluate a one-off course, now completed and never to be run again? We might, but the course itself is not the only variable which can be changed. Should the institution support similar courses in the future? Should clients purchase similar courses or seek alternatives? Should staff change their approach to the design or presentation of courses? If no question of change arises then evaluation is deprived of point and becomes instead a public relations exercise service to induce feelings-of-being-consulted in the course consumers, or feelings-of-being-valuable in the course presenters. Both of these goals can be achieved more effectively without an evaluation charade. Evaluation and change are logically and methodologically linked in a fundamental way. Just as 'facts' in science only become 'evidence' when they are ranged for or against some theory, so course data become evaluation data only in the light of contemplated change. There are a number of reasons why this relationship between change and evaluation deserves emphasis:

- External validation in PSHE has familiarized institutions with an evaluation process in which their own role has been largely confined to passively displaying data. For administrative reasons data generation (institutions) and evaluation decisions (CNAA, etc.) have been artificially stratified. There is a serious danger that post-accreditation validation procedures will simply internalize this same separation within institutions.
- The bulky results of separating data generation from evaluation decisions have for years been a burden to teaching staff, a drain on institutional resources and a crime against the arboreal environment. Less obviously, the separation is a methodological disaster which left to themselves, institutions had no reason to develop, and now have no reason to perpetuate. In science it is impossible to proceed by listing all relevant facts, then locating an appropriate theory, because the list of facts is usually unmanageably long. Indeed, there is no way of defining 'relevant', and thus maintaining the list within finite bounds until a theory is specified. Instead scientists proceed iteratively, arriving at an alternative theory set on the basis of a restricted fact base, and then

Table 6.2 Five years of Modular Course evaluation

Date	*Brief description of project*	*Method(s) employed*	*Personnel*
June 1982	Student impressions of Modular Course	Questionnaire	EO+Dean
October 1982	Attitudes of finalists to Modular Course	Structured interview	EO+Dean
July 1983	Lancaster Approaches to Studying Questionnaire	Ratings/survey $N=241$	EO+EMU+ external consultants
June 1984	Replication of approaches to studying	Ratings/survey $N=192$	EO+EMU
June 1984	Modular Course management structures	Documentation analysis	EO
June 1984	Employability survey	Interview/ survey	EO+ graduate research assistants
May 1985	Gender and recruitment	Secondary data analysis	EO+student project
July 1985	Gender and bias in project assessment	Secondary data analysis	EO+ Psychology Unit
January 1986	Contact hours and study strategies	Secondary data analysis	EMU+ Geography Unit
January 1986	Cooperative and communications skills	Survey of Field Chairs	EO
January 1986	Workload variability	Questionnaire/ rating scales	EO+EMU
January 1986	Resource changes and academic performance	Secondary data analysis	EO+ Psychology Unit
May 1986	Academic performance in Stage I	Secondary data analysis	EO+ Psychology Unit
May 1986	Module timetabling	Rating scale/ questionnaire	EO+EMU
July 1986	Public images of higher education in universities and polytechnics	Rating scale/ questionnaire	EO+project student
January 1987	Need for institutional change	Interview/ rating scale/ survey	EO

generating facts which are maximally likely to discriminate between theories. This is the model which is most appropriate for evaluation activities within higher education institutions.

- A decision that change is necessary is rationally justified when there is an unacceptable difference between a current state and a desired state. It follows that evaluation evidence depends for its existence and interpretation upon a specification of desired states. Emphasis upon the link between evaluation and change draws attention to the often neglected fact that evaluation evidence cannot determine system goals nor can it indicate whether system goals are being optimally achieved unless the goals are clearly specified. In British higher education the precision with which goals are stated is usually inversely related to their institutional generality; thus course and grassroots teaching teams usually have clearly articulated goals; faculty goals are often pious but hopelessly vague; institutional goals are rarely stated and are never sufficiently defined to be operationalizable. This contrasts strongly with higher education in the USA where institutional 'mission statements' are regarded as selling points, and key senior administrative staff may be hired to further achievement of specific goals; and fired for demonstrably not doing so. Ball and Halwachi (1987) have recently argued that performance indicators cannot be effectively employed within British higher education because of the absence of clearly specified institutional goals. Within Oxford Polytechnic, one consequence of Modular Course evaluation activities has been to sharpen thinking about course and institutional goals.

Evaluation of the Modular Course

The shortcomings of monitoring and review procedures are one reason why other forms of evaluation operating alongside them are desirable. New possibilities permitted by modular structure are a second. For almost the first decade of its existence, the resources of the Modular Course were entirely absorbed by coping with the first-order problems of survival and growth. By the early years of the 1980s, there was sufficient breathing space for the second-order problem of self-improvement to be addressed. The earliest steps in this direction were taken by David Watson, the first permanently appointed Dean of the Course. One of Watson's main contributions to the development of the Modular Course has been to build in evaluation mechanisms at many levels, to foster positive attitudes to evaluation by using evaluation data in course development and not to shrink from its application to his own roles and activities.

The early studies were wide angle studies of impressions of the Course, based around questionnaires and structured interviews. There was some nervousness that perhaps modular courses did not deliver a satisfactory educational experience from the viewpoint of the student. As confidence

Table 6.3 Seven years of feedback seminars

Term	*Topic*
Spring term 1982	Successful models of course evaluation
Summer term 1982	Evaluation of fieldwork and fieldcourses
Autumn term 1982	Improving employability
Spring term 1983	Employability – the video (showing of in-house video production, with comment from invited external speaker)
Summer term 1983	Designing module evaluation questionnaires
Autumn term 1983	Integration and interdisciplinarity on the Modular Course
Spring term 1984	Students' and supervisors' views of dissertations
Summer term 1984	Recruitment to the Modular Course
Autumn term 1984	The mature student on the Modular Course
Spring term 1985	Assessment bias and its control
Spring term 1985	Organizational culture in credit accumulation systems
Summer term 1985	Novel forms of assessment
Autumn term 1985	Staff perceptions of the Modular Course
Spring term 1986	Teaching in 4-hour blocks
Summer term 1986	Uses and abuses of external examiners
Autumn term 1986	'Vocational' versus 'academic' in higher education
Spring term 1987	Non-traditional teaching
Summer term 1987	Staff morale
Autumn term 1987	Independent study
Spring term 1988	Flexible use of teaching space
Summer term 1988	Personal tutoring

grew, partly as a result of the student impression studies, so techniques were employed which permitted cross-institutional comparison, which enabled graduate employer samples to participate in evaluation, and which allowed the appraisal of course realities such as grading patterns and resources to be set alongside the studies of course perceptions. An impression of the nature and pace of these changes is given in Table 6.2.

These developmental changes in perspective and technique were deliberate. The administrators of the Modular Course, particularly Watson, realized early that sustainable course quality depends crucially upon relevant feedback and that course development strategies based upon relevant evidence receive readier acceptance from staff than strategies deriving their credibility from the perceived wisdom of management. Accordingly a series of steps were taken which aimed to establish an academic culture within which evaluation activities were widespread and respected; evaluation skills and expertise were readily available; and evaluation gradually became an expected element in the body of knowledge and skills possessed by professional higher education practitioners. These steps were:

- top-down initiation of evaluation studies;
- dissemination of knowledge and skills by:
  1. establishment of an ongoing course-wide series of termly seminars dedicated to evaluation activities (Table 6.3 gives a list of evaluation issues covered by these seminars during the last seven years);
  2. establishment of a termly newsletter, *Feedback*, dedicated to evaluation issues and news, and distributed free to all staff; and
  3. appointment of a Modular Course Evaluation Officer (EO).

The first Evaluation Officer was appointed in September 1982. The EO was a 0.3 FTE two-year post, the appointment being made from among staff on the Course. It was financed by a levy across all departments contributing to the Modular Course. The duties of the EO were described in the 1983 Modular Course *Staff Guide* as follows:

> The appointment of the Modular Evaluation Officer ensures that the broader aspects of the Course are (also) reviewed. The Evaluation Officer provides leadership and coordination for the evaluation of the Course and is the editor of *Feedback*.

The main job of the EO was to foster an evaluation culture. But like the risen Christ, he had also to prepare the faithful for his disappearance from the world. It was hoped that after very few incumbencies evaluation activities would be sufficiently deep-rooted to be self-sustaining, and the post could be abolished.

A second task for the EO was to develop a model of Modular Course evaluation, which related possible objects of evaluation to available techniques and the results of evaluation exercises to the mechanisms of course administration and development. This overall model of Modular Course evaluation has to be definable within the framework of measurement theory and good evaluation practice; but it must also take account of the needs and distinctive features of the Modular Course. The main constraints from measurement theory are summarized below. Constraints imposed by good evaluation practice are rather more severe: an elegant statement of them was produced by the US Joint Committee on Standards for Educational Evaluation (1981). This statement is reproduced as Table 6.4. All Modular Course evaluation studies have attempted to operate within the Joint Committee standards.

There have been distinct evolutionary changes in the focus, style and methodology of evaluation. Table 6.2 summarizes major evaluation projects over the five years from 1982–87. Changes in style are in part due to the emphases of different Evaluation Officers. The post was designated as a rotational one to ensure periodic changes in theoretical outlook. Changes in focus have resulted partly from incorporation of some evaluation

Table 6.4 Summary of standards for evaluations of educational programmes, projects, and materials

A	*Utility standards* The Utility Standards are intended to ensure that an evaluation will serve the practical information needs of given audiences. These standards are:
A1	*Audience identification* Audiences involved in or affected by the evaluation should be identified, so that their needs can be addressed.
A2	*Evaluator credibility* The persons conducting the evaluation should be both trustworthy and competent to perform the evaluation, so that their findings achieve maximum credibility and acceptance.
A3	*Information scope and selection* Information collected should be of such scope and selected in such ways as to address pertinent questions about the object of the evaluation and be responsive to the needs and interests of specified audiences.
A4	*Valuational interpretation* The perspectives, procedures, and rationale used to interpret the findings should be carefully described, so that the bases for value judgements are clear.
A5	*Report clarity* The evaluation report should describe the object being evaluated and its context, and the purposes, procedures, and findings of the evaluation, so that the audiences will readily understand what was done, why it was done, what information was obtained, what conclusions were drawn, and what recommendations were made.
A6	*Report dissemination* Evaluation findings should be disseminated to clients and other right-to-know audiences, so that they can assess and use the findings.
A7	*Report timeliness* Release of reports should be timely, so that audiences can best use the reported information.
A8	*Evaluation impact* Evaluations should be planned and conducted in ways that encourage follow-through by members of the audiences.
B	*Feasibility standards* The Feasibility Standards are intended to ensure that an evaluation will be realistic, prudent, diplomatic, and frugal; they are:

Table 6.4—continued

B1	*Practical procedures* The evaluation procedures should be practical, so that disruption is kept to a minimum, and that needed information can be obtained.
B2	*Political viability* The evaluation should be planned and conducted with anticipation of the different positions of various interest groups, so that their cooperation may be obtained, and so that possible attempts by any of these groups to curtail evaluation operations or to bias or misapply the results can be averted or counteracted.
B3	*Cost effectiveness* The evaluation should produce information of sufficient value to justify the resources expended.
C	*Propriety standards* The Propriety Standards are intended to ensure that an evaluation will be conducted legally, ethically, and with due regard for the welfare of those involved in the evaluation, as well as those affected by its results. These standards are:
C1	*Formal obligation* Obligations of the formal parties to an evaluation (what is to be done, how, by whom, when) should be agreed to in writing, so that these parties are obligated to adhere to all conditions of the agreement or formally to renegotiate it.
C2	*Conflict of interest* Conflict of interest, frequently unavoidable, should be dealt with openly and honestly, so that it does not compromise the evaluation processes and results.
C3	*Full and frank disclosure* Oral and written evaluation reports should be open, direct, and honest in their disclosure of pertinent findings, including the limitations of the evaluation.
C4	*Public's right to know* The formal parties to an evaluation should respect and assure the public's right to know, within the limits of other related principles and statutes, such as those dealing with public safety and the right to privacy.
C5	*Rights of human subjects* Evaluations should be designed and conducted, so that the rights and welfare of the human subjects are respected and protected.
C6	*Human interactions* Evaluators should respect human dignity and worth in their interactions with other persons associated with an evaluation.

Table 6.4—continued

C7	*Balanced reporting* The evaluation should be complete and fair in its presentation of strengths and weaknesses of the object under investigation, so that strengths can be built upon and problem areas addressed.
C8	*Fiscal responsibility* The evaluator's allocation and expenditure of resources should reflect sound accountability procedures and otherwise be prudent and ethically responsible.
D	*Accuracy standards* The Accuracy Standards are intended to ensure that an evaluation will reveal and convey technically adequate information about the features of the object being studied that determine its worth or merit. These standards are:
D1	*Object identification* The object of the evaluation (program, project, material) should be sufficiently examined, so that the form(s) of the object being considered in the evaluation can be clearly identified.
D2	*Context analysis* The context in which the program, project, or material exists should be examined in enough detail, so that its likely influences on the object can be identified.
D3	*Described purposes and procedures* The purposes and procedures of the evaluation should be monitored and described in enough detail, so that they can be identified and assessed.
D4	*Defensible information sources* The sources of information should be described in enough detail, so that the adequacy of the information can be assessed.
D5	*Valid measurement* The information-gathering instruments and procedures should be chosen or developed and then implemented in ways that will assure that the interpretation arrived at is valid for the given use.
D6	*Reliable measurement* The information-gathering instruments and procedures should be chosen or developed and then implemented in ways that will assure that the information obtained is sufficiently reliable for the intended use.
D7	*Systematic data control* The data collected, processed, and reported in an evaluation should be reviewed and corrected, so that the results of the evaluation will not be flawed.

Table 6.4—continued

D8	*Analysis of quantitative information* Quantitative information in an evaluation should be appropriately and systematically analysed to ensure supportable interpretations.
D9	*Analysis of qualitative information* Qualitative information in an evaluation should be appropriately and systematically analysed to ensure supportable interpretations.
D10	*Justified conclusions* The conclusions reached in an evaluation should be explicitly justified, so that the audiences can assess them.
D11	*Objective reporting* The evaluation procedures should provide safeguards to protect the evaluation findings and reports against distortion by the personal feelings and biases of any party to the evaluation.

Source: The Joint Committee on Standards for Educational Evaluation, McGraw-Hill, 1981.

functions such as course perception sampling into routine mechanisms of modular and field management. Another reason for changing focus has been a shift away from evaluation data as adequacy assurance towards evaluation data as a source of developmental information. This has naturally been reflected in changes in methodology, particularly in a movement away from course impression analysis towards student performance analysis.

Methods of evaluation

Evaluation methods are a subset of measurement procedures. Measurement is the systematic mapping of a variable on to a set of descriptors. These descriptors may be words such as 'good', 'bad', etc., or numbers. All measurement procedures are associated with bias (responsivity to variables other than that being measured) and reactivity (interference with the value being measured by the measurement procedure). In educational measurement bias and reactivity are hard to estimate but rarely negligible. The pragmatic implication of this is that whenever possible more than one technique should be employed.

There is a relevant distinction here between data that are independent of the evaluator's own judgements and data that are not. Evaluation data are obviously never value free; but even subjective data such as student evaluation of courses can be recorded objectively, e.g. self-recorded written responses can be analysed by computer or an independent third party; or subjectively, e.g. a request for comment from the class can be written up by the course tutor three weeks later.

Subjective methods

Illuminative observation

Perhaps the best known subjective evaluation procedure is the 'illuminative observation' technique (Parlett and Hamilton, 1972). Essentially this consists of a series of extended interviews, together with a scrutiny of documentation and similar deliverables, resulting in a report in which conclusions are illustrated by quotations from interviews, or descriptions of what has been observed. The utility of such reports obviously depends upon the calibre and acumen of the observer, the typicality of the materials studied and the sample interviewed. The technique could more illuminatingly be called 'impressionistic observation'. It is obviously very prone to bias and reactivity effects; it can easily be subverted by stacking samples; and the credibility of the report depends completely upon the creditability of the practitioner. As a consequence it can be difficult to make unpopular recommendations stick: either the credentials of the investigator or the weight of the data can easily be challenged by those with a will to do so. A case in point is a study commissioned by Oxford Polytechnic's Modular Course. The focus of the study was the final year honours project module and the investigator was a consultant of considerable experience, from outside the Polytechnic. In the evaluation report some wise and many sensible things were said. However the effect of the report was determined entirely by reaction to the 'fear and apprehension . . . subculture' (*Feedback*, 1984, p. 3–4) which the investigator claimed to have detected. Though the final report was dutifully disseminated by the course administration, it was greeted with great scepticism by staff and rapidly forgotten. The reason is that the vivid phrase which the investigator had chosen to 'flag' what he considered to be an issue of concern was seen as exaggerated and alarmist by academics. They were only too used to the need for chivvying insouciant students into beginning some work for the module as deadlines approached. Once a single strand of the report was challenged, its whole credibility was unravelled: the 'subculture of fear' conclusion was not proven, but only illustrated by the report; there were no independent data to show that a general problem existed. It was thus easy, and probably correct, to write the alleged anxiety problem off as a failure of the investigator's sense of proportion. But if in this, why not elsewhere? Once credibility is challenged it is difficult to justify selectively.

In other contexts, the illuminative observation approach has been used by Oxford Polytechnic with more success. A central service called the Educational Methods Unit may be called upon by academic teams to provide independent evaluations of courses or to advise upon specified problems. Illuminative observation is often used and generally gives good results. The salient differences between these two applications are that in the second case the consultant, though independent of the course team, is known to be supportive and familiar with the background and context of the course under examination; the quality of the consultant's judgement is also known to most of the course team. It seems a sensible conclusion that the strengths of this technique lie not in the method itself, but in the experience and skill of the person who employs it, and the extent to which these are known and respected by the beneficiaries of the study. Unpalatable conclusions are likely to require more substantial evidence than this technique can provide.

Interviews
Interviewing is the subjective technique which is most widely used in higher education. Early evaluation efforts on the Modular Course used interviews to investigate student impressions of the Course (1982; see Table 6.2). Interviewing is vulnerable to reactivity and bias effects as well as sampling difficulties. Whenever the validity of interviewing as a decision support technique can be appraised, it is very poor. Unfortunately it is also associated with an 'illusion of competence' effect which leads those employing interviews to value the quality of their decisions out of all proportion to their probable true worth. It would be rash to conclude that interviews have no role to play in evaluation. They do provide a rapidly accessed source of data and they can be a fruitful source of issues and hypotheses which should be explored in other ways. The face-to-face interaction feature of interviews which makes them such a potent source of interviewer bias also makes them an effective device for communicating concern, interest and priorities to interviewees. The other side of the reactivity coin is that change induced in respondents by evaluation may be of positive benefit. However, interviews should rarely be used as a sole basis for evaluation decisions of any great significance. When interviewing is employed, the use of multiple independent interviewers is a sensible precaution. Validity is also likely to be enhanced if the interview is recorded and transcribed before analysis. Techniques such as content analysis can make the analysis less impressionistic. The use of 'structured' interviews, with the same items being used for all interviewees, allows the generality of responses to be evaluated. (But see the next section for implications of this.)

Objective methods

Questionnaires and rating scales
In terms of form and content a questionnaire is a written version of a completely structured interview. The main differences are in terms of the social processes accompanying inquiry. As a result, the investigator is not exposed to a number of major sources of bias, such as physical appearance and accent. Reactivity is also likely to be reduced as respondents can easily remain anonymous – impossible with interviews except when external consultants operate in very large institutions – and there is less social

pressure to produce responses which are agreeable to the investigator. The disadvantages are that informants are less likely to respond, or to respond seriously, and that the investigator is not able to follow up unanticipated developments.

These disadvantages can usually be coped with providing they are recognized. Low response rate may actually be genuinely signalling a lack of concern about the evaluated issue, a fact which can be masked by interviews in which respondents feel constrained to provide appropriate positive and negative responses when asked. In some contexts it may be appropriate to educate a population on the importance of response, to administer the questionnaire to a 'captive group' or to offer modest incentives.

Loss of opportunity to respond to the unexpected can be dealt with only by striving to eliminate it. In practice this means constructing questionnaires on the basis of preliminary interviews, administering pilot versions of the questionnaire under conditions which allow respondents to identify shortcomings and so on. If precautions are taken to minimize the weaknesses of questionnaires they can be quick and convenient to administer, easy to analyse and have higher validity than more time-consuming alternatives such as interviews. This is not to say that validity should be taken for granted: McBean and Lennon (1985) have shown that though for large groups, response rates of 50 per cent are adequate; with groups of less than 30, an 80 per cent response rate is required to give course ratings which are within 12 per cent of the mean for the population as a whole.

Some studies have raised even more fundamental questions about the validity of course evaluation questionnaires. Moses (1986), among others, has found that student ratings of lecturer performance were entirely unrelated to self-ratings by the same staff. Miron (1988) has reported, for an admittedly small university sample (93 classes), that self-ratings of course presentation by newly recruited staff are significantly closer to student ratings than are the ratings of older staff. Miron's tentative conclusion is that older staff may become less concerned about teaching as they develop other priorities; but it is equally possible that students are responding more similarly, and more favourably, to lecturers who are more like themselves.

Less controversially, it soon becomes apparent to personnel involved in course and presenter evaluation that student respondents rarely assign very poor ratings, or make extremely negative comments, even when other indicators such as colleague judgement suggest that a course has been presented particularly poorly. This does not impugn the value of student ratings as a basis for relative judgements; but it does suggest that caution should be exercised in making absolute interpretations.

It would be disastrous to morale to introduce a high-powered multiple-indicator methodology in order to isolate which staff really are poor performers and which above average. There is a sense in which students cannot be completely wrong. If a lecturer is not producing an experience which students regard as satisfactory, even when academic assessments are fine, some aspects of that lecturer's performance deserve attention. The pragmatic route through this hazardous thicket of management and evaluation issues which has evolved on the Modular Course has four elements:

- Staff are expected to obtain student evaluations of a module whenever it runs. These evaluations may remain confidential and are expected to guide self-development.
- Students are given opportunities at field and Modular Course level to raise difficulties not addressed at the level of modules. All fields must identify field representatives, elected by students from among themselves, who are expected to bring problems within a field to the attention of the Course administration.
- The three criteria for promotion within the institution require the demonstration of excellence in teaching, research and administration. Advisory literature indicates that summaries and illustrations of student evaluation are an important source of evidence with regard to the first criterion.
- The Polytechnic staff development policy requires that each lecturer shall have an annual opportunity for a development interview with a Head of Department. Problems and progress in the evolution of teaching skills are an important agenda item for such meetings.

Thus an attempt is made to harness data from course perception questionnaires to assist in the self-development of staff. This use assumes the importance of student impressions, but not their absolute truth; it offers opportunity and incentives for self-development, but no direct penalties; and it provides a mechanism for students to blow the whistle if problems move outside the range of acceptability.

The difficulties associated with questionnaires are usually also applicable to rating scales. Rating scales are a subset of questionnaires which invite informants to select a number corresponding to the extent of their response, or to select one of a series of descriptions, usually corresponding to an ordinal scale with a neutral midpoint, to which the investigator can easily assign numbers. A common example of such a series of descriptions is:

Very unsatisfactory;
Unsatisfactory;
Indifferent;
Satisfactory; and
Very satisfactory.

The main advantages of rating scale data are that responses are usually simple and uniform and easy to code, transform, statistically analyse and describe. These advantages are such that ratings are to be preferred when they can be used without distorting the intentions of the investigation.

Some examples which illustrate successful and unsuccessful uses of these methods on Modular Course evaluation projects may be helpful. In the Modular Course Employability Survey (1984; see Table 6.2) the problem of respondent motivation was addressed by:

- choosing the evaluation project on the basis of a group discussion with randomly selected students who were asked to indicate, and justify, what they would like to see evaluated;
- employing personnel who had graduated from the course within one year to carry out interviews; and
- using the same graduate assistants to administer the rating scales to student respondents personally.

Aspects of the Course to be evaluated by employers and students for contribution to employability were selected on the basis of preliminary interviews with staff and students, but supplemented and refined after a pilot questionnaire administered to a small random sample of students (n=20) had been analysed. When the main scale had been constructed, student and employer samples were asked to select one from a range of scaled descriptive phrases indicating the employability value of different features of the course. Because the data were easily convertible into numerical form, it was a straightforward exercise to seek correlations and larger than chance differences in perception between first and third year students, and student and employer samples. The completion rate for rating scales in this study was 100 per cent for the student samples. This is exceptional: response rates of 10 per cent and under are not unusual if postal responses are required. But response rate is to some extent within control: a Modular Course survey of staff satisfaction with modification of a scheme for mapping courses on to timetable hours, yielded an (internal) postal return rate of 35 per cent. The high return rate here was probably due to brevity of the survey form – only six sample responses were required – and again, to choice of issue. Unlikely as it may sound, the timetabling scheme aroused strong feelings among staff.

Secondary analysis of data

Qualitative data

Multidisciplinary modular courses require considerable and continuous effort to be devoted to harmonizing and integrating the practices of academic components. This inevitably produces copious documentation. Documentation must be as explicit as possible because staff and circumstance seem to take a rabbinical delight in producing difficult cases. This documentation is itself a rich source of material for evaluation. It does not suffer from the bias factors operating when documentation is generated, not to regulate practice but to inform reviewers. One example of how course documentation can be invaluable for evaluation was a documentation analysis carried out on Modular Course *Staff Guides* and course resubmission documents in 1984 (see Table 6.2, Modular Course management structures).

The analysis was commissioned by the Dean, in response to concern from members of the MCC that the management and course development role of the Committee was seriously diminishing. Field Chairs are not paid for their additional responsibilities, nor do they automatically receive any remission of teaching (though most fields have developed arrangements which partially compensate for the considerable demands of the post). The most important incentives for Field Chairs are thus the opportunity to exercise some control over the educational context within which their subject is taught; the challenge of defending the interests of their discipline; the opportunity to develop expert knowledge of the course structure and regulations through participation in debate; and the advantage of administrative experience for career development.

The central administration of the Modular Course has always been keenly aware of the burden falling on Field Chairs. It also tended to regard the Modular Course Committee (MCC) as excessively large and difficult to manage. Consequently, and for thoroughly laudable reasons, there had been a systematic drift of responsibility from MCC to its executive subcommittee, MMRC, between 1979 and 1984. The number of meetings of MCC, for example, had diminished by half, its published responsibilities had altered, and even its relationship to MMRC in the *Staff Guide* management system flowchart had altered. None of these changes were deliberate, but none had been approved by MCC. A comparative analysis of course documents made the extent and systematic nature of the changes quite evident.

The unintended consequences of the diminished role of the MCC were to erode the very functions of Field Chairs which made the post worthwhile, to make them less informed by removing opportunity for MCC debate and replacing it with reports of MMRC debates, and to reduce the contribution that fields could make to Modular Course development.

The main conclusions of the 1984 study of management structures were debated and accepted. The result was a reaffirmation of the status of MCC as a senior committee with respect to MMRC; a recognition that changes in Course regulations must receive approval by MCC before appearing in course documents; and restoration of the original frequency of MCC meetings. Recognition of the importance of Field Chair contributions to course evolution was achieved by designating one MCC meeting per term as a development meeting. This innovation provided, for the first time, an explicit mechanism for preparing for the future, in addition to responding to present issues. It also provided a point in the system at which results of evaluation activity could receive appropriate consideration.

Quantitative data

By their very nature modular courses generate large amounts of quantified or easily quantifiable data. The primary purpose of these data is to record assessment decisions, to support decision-making by management, or to enable monitoring of quality to occur. Much of these data, however, are also suitable for hypothesis-testing – indeed the data are of higher quality than those generally used in research studies because it is not usually practicable to ask academics from a wide range of areas to adopt uniform teaching, assessment and reporting practices solely to permit educational research to occur. One simple example illustrates this point. The example concerns sex bias, which is of particular interest to the Modular Course at Oxford Polytechnic because of the high proportion of female applicants which it attracts. In fact the relationship between gender, application rate and admission probability has itself been the subject of an evaluation study (1985; see Table 6.2).

In 1984 Clare Bradley of Sheffield University published a persuasive paper (Bradley 1984) arguing that sex bias affected the assessment of student projects. This bias was detectable when assessors were relatively unfamiliar with the students being assessed, and were aware of gender information through use of first names. The bias effect operates to increase the spread of marks for males relative to females. Its consequence is that grades of both high-performing women and low-performing men are depressed compared with their opposite gender peers. Dr Bradley was invited to present her findings to a Modular Course seminar; they were publicized in a widely disseminated evaluation newsletter. As a result, it was decided to:

- shift immediately to an initials only format on the title page and assessment blank for student projects;
- analyse project marks retrospectively to seek evidence of gender bias; and
- reanalyse marks after the change in name style to seek any consequent change in mark distributions.

All of these procedures could be carried out using data which was generated as a matter of course. No gender differences were in fact detected, but the assessment system was rendered less liable to such biases, staff were sensitized to their existence and nature and the course was able to give itself a clean bill of health without any special data collection effort being required.

On a much more ambitious scale, Lindsay and Paton-Saltzberg (1987; see Table 6.2) analysed data from 36,984 student assessments in arts, social studies, science and applied areas, to show that the frequency of high grades decreases and that of lower grades increases as the number of students enrolled on a module increases. This effect occurred over all four academic areas in spite of consistent and striking assessment differences between them. Apart from providing important evidence about the relationship between resources and performance in higher education, the demonstration of stable between-area assessment patterns has enabled the Modular Course to successively nudge each area into modifications of assessment practice which gradually bring them closer together. Similarly, the finding that class size affects performance, and ancilliary evidence on the rate of increase in the frequency of large classes, has directly led to the provision of additional purpose-built accommodation for large classes. A continuing programme to investigate, evaluate, and disseminate best practice information on teaching large classes has also been developed.

A similar study of modular Stage I assessment data (1986; see Table 6.2) showed that there was imbalanced modular provision and enrolment over the academic year. Most strikingly, fully 57 per cent of module failures occurred in term 3, even though only 38 per cent of assessments occurred in that term. This finding was relatable to Stage I regulations which attempted to encourage students to broaden their educational horizons by requiring them to take 12 modules, but to pass only 10. The practical effects of this regulation were to cause many students to register for 12 modules but participate seriously in only 10; to cause them to stack their choices, so that modules which they needed or intended to pass were taken in terms 1 and 2, and to bias apparent but not real demand towards term 3. Once these effects were demonstrated, resistance, which had been spirited, to changing the Stage I regulations to a 'take and pass 10' formulation, tended to dwindle away; few were motivated to defend the educational benefits of phantom registration.

Secondary analysis of data has been an extremely useful source of evaluation evidence. It is cheap and easy to collect and use. It is necessarily of the same quality as the data used to support managerial control, because it is that data, used for a different purpose; and because the data is of a type already in currency, its use needs no special explanation or defence.

Presentation of evaluation results

Presentation of the results of evaluation has proved over and over again, in Oxford Polytechnic's experience, to be one of the most critical variables in determining whether appropriate change is produced. In general, evaluation research tends to have low credibility among academics. Consequently the results of each study have to be sold on the basis of the appropriateness of the methodology employed to the problem addressed. One implication of this is that evaluation procedures are usually better developed in-house than bought in from other institutions. An example was the use of the Lancaster Approaches to Studying Inventory which was administered to two large samples of Modular Course students in 1983 and 1984 (see Table 6.2). The theoretical assumptions of the Lancaster Inventory are considerable, and did not appear to be widely shared by staff on the Modular Course. As a consequence it was impossible to draw any practical conclusions from the results of this evaluation exercise.

A second implication of the low background credibility of evaluation in higher education is that the methodology of each study must be fully described. The Modular Course has gone about this by asking members of the evaluation team to present an account of each major evaluation study and its results to MMRC and MCC, the two senior Course management committees. This is usually accompanied either by a full written report, or by a brief report with a fuller version lodged with each departmental office, if the complete report is lengthy. An account of the study and its main findings is separately written for a termly evaluation newsletter which is distributed to all staff. A further presentation of the study will usually be incorporated into a termly seminar series which focuses on evaluation and course development issues. All interested staff are thus exposed to the result of each major study, and get a chance to query or challenge on points which they do not immediately accept. This strategy is not without its dangers. Because methodology is fully exposed, genuine defects may be revealed, or more frustratingly, discussion sidetracked away from implications. The experience of the Modular Course is that the risks are worthwhile, because when the evaluation study does carry conviction, consequential course changes are accepted as a natural path for course development to take given the known facts, rather than the latest wheeze of course management.

The use of statistics has proved to be a particularly thorny presentation issue. Statistical techniques are often essential for fully understanding a complex data set. In presenting results, however, statistical concepts and terminology can easily alienate or distract. For example, in the Modular Course evaluation study of 'workload variation' in coursework between modules and fields (January 1986; see Table 6.2) rating scale responses were subjected to factor analysis. The results were of considerable interest: 'surface workload' was clearly distinguishable from a 'cognitive stretching' factor related to quantity of new concepts, and assessment load for example. 'Underload', associated with a poorer idea of progress, uncertainty about how to deploy time, panic at exams, etc., was perceived almost as negatively as overload. A general lack of understanding of factor analysis, however, weakened the impact of the study and its effect was minimal.

Similarly, in the need for institutional change study it was clear that need for change has two distinct components: the deficiency level of some institutional subsystem, and its importance to the system as a whole. To give an example: a leak in the roof of one's garden shed would probably be seen as less in need of urgent action than exactly the same defect in the roof of one's home or car. This distinction was handled by asking respondents to rate each of two hundred features of the Modular Course and its institutional context, both for importance and for need for change. Data were then presented as rank-ordered products of the mean ratings. It was predictable that the onerousness of the rating task held down response rate somewhat (33 per cent of a target population of 200), but less predictably, great suspicion was aroused by the use of multiplication as a way of combining the two components of need for change. Again, potentially valuable data were largely wasted because of a mistake in presentation. The moral is that the need to clearly communicate findings must always take precedence over considerations of technical adequacy.

Conclusions

In an environment as diverse and complex as Oxford Polytechnic's Modular Course, monitoring and review processes are a crucial element in maintaining course quality. Both processes constantly tend to become perfunctory, routinized and burdensome. This tendency has to be limited by a number of mechanisms which the course has evolved. Monitoring and review alone are not sufficient to ensure optimal course quality. Problem-centred evaluation studies using a range of methods are a necessary supplement. The focus and methodology of such studies tend to shift as the course matures. Successful evaluation techniques often become absorbed into the routine processes of course management. Secondary analysis of routinely generated data has proved particularly valuable on the Modular Courses. The manner in which the results of evaluation studies are presented has a considerable effect on their impact.

7

Systems

Chris Coghill

The record-keeping and office procedures implications of the Modular Course when it started in 1973 with an intake of 75 students were modest but have since grown with the Course. There is now the need to keep track of thousands of individual student module programmes, termly timetables and assessments and of applications for places across hundreds of field combinations. This growth would have been strangled by the paperwork involved if new computerized systems had not been introduced and continuously developed.

The foresight of the Course founders, in using a computer for a range of student records applications from the very first intake, set the pace for subsequent information technology developments. That original range was surprisingly wide and included personalized student records,[1] timetables and class lists.

However, by 1979 the computerized course support was staggering under the load. Data files included punched cards (as data files, not just for input); the computer was dated, overloaded and likely to be replaced by one from a different manufacturer; there was little real-time access to data; management information was scarce; there was little in the way of computerized records for Polytechnic students on non-modular courses, and that little was unintegrated with the modular system; there were no benefits or access to data for offices besides the registry; the file structures could no longer cope with the then size of the course (1,100 students); and the data structures were arcane (for example, over-zealous use of bit-packing). Documentation, training and the involvement of junior office staff in the running of such a system were understandably difficult to achieve.

[1] Throughout, the term 'student record' (sometimes 'record') refers to the summary progress document prepared for students each term (and for the examinations committees each vacation) showing personal details, modules taken and results, those being taken and to be taken, and summary of progress *vis-à-vis* the regulations. Sample records are given in Chapter 1.

To remedy these defects a new *student management system* was designed and introduced in stages from 1980–81. The design, and all subsequent developments, were guided by the following principles:

- to meet the needs of a very wide range of offices and individuals, including:
  students;
  teaching staff, admissions tutors;
  registry, modular office;
  departmental and faculty offices;
  library;
  student services;
  finance office;
  examinations;
  course management;
  directorate.
- to meet the management information needs of the institution as a whole through the seamless integration of modular and non-modular course records, together with the needs of external bodies;
- to be flexible in design and use in order to meet new demands for information from within and without the Polytechnic, to adapt to growth and structural change (within both courses and the institution), and to answer effectively any *ad hoc* enquiry (this flexibility is also required by the method of project management described below);
- to develop from a small beginning in manageable steps rather than to spend a long period identifying 'all' institutional needs and all required data items followed by software development of a 'complete' system on an unmanageable scale;
- to be designed around the established needs of the Course and its working practices, academic structure and regulations, in order to assist all staff and students to fulfil their roles (on occasion, analysis for proposed developments showed up imperfections in existing procedures and sharpened user perceptions of their roles and working practices);
- to put contol in the hands of the users. This means not only office staff, who bear the brunt of system usage, but teaching staff, who must retain specific decision-making roles, and, last but not least, the students themselves, who must have faith in the system's accuracy and ability to provide records and timetables of appropriate quality at specific times and on demand;
- to be designed, developed and maintained to the highest achievable standards of data processing, subject only to resource constraints; and
- not to attempt to solve problems that do not exist or whose solution rests with human reorganization.

The student management system in practice

Each *student* is held responsible for registering his or her own module programme. To do this effectively requires, besides the array of guides and handbooks and the various counsellors (tutors, field chairs, etc.), well designed forms (for both input and output) and efficient data processing.

Having registered his/her programme a student may request, at almost any point, a freshly printed record (normally available within 24 hours). In addition, an updated record is provided for each student following the termly examination committees, showing the student's results. An updated timetable is provided at the start of each term, showing the days, hours and rooms for each registered module. Both records and timetables will contain messages and instructions specific to individuals if invalid module registrations have been attempted.

These documents – student records and timetables – often have to carry a great deal of information including complex messages and progress summaries. Their design, together with the content and format of the printed data, has received particular attention. The ergonomics of form design and use have had a pivotal role in the provision of accurate, presentable and punctual information to all users. The attention to detail has covered paper sizes, colours and qualities, forms handling machinery, punch holes for ringlock binders, minimal use of codes, preprinted marginal notes, and devices to aid visual impact such as tints, contrasts, colour, corner flashes, etc.

On leaving the Course, any student who has passed at least one module will receive a glossy covered printed transcript showing all the modules passed, their credit value, the grades achieved, the fields (if any) studied, and the award (if any) made. Transcripts are also available on demand to support any student in job hunting while still on the Course.

Teaching staff have a number of needs for information, and for assistance with recording assessment data and keeping other course records. In the termly cycle, a module leader will receive a succession of reports, all with variable selections and sort orders, at predetermined points and on demand. These reports include:

Module timetable
confirming the agreed times and groups and detailing which rooms have been booked.

Class lists
showing all students registered on the module, their set (or timetabled group), their other modules, fields, personal tutor.

Star charts
showing the *untimetabled* hours during the week for each student on a module; used for arranging tutorials, alternative class meetings, etc.

Coursework record sheets
containing the names of all students on a module, with printed columns for the entry of coursework assessments.

Marksheets
on which the final assessment for each student on the module is entered, for subsequent data processing by the registry.

In addition, personal tutors receive copies of their tutees' records and timetables, and admissions tutors receive regular and on demand analyses of applications data to support their decision making and to assist the meeting of both individual field and overall Course intake targets.

The Registry and its organization is central to the effective management of information. Registry staff must be able to process information without introducing delays, whether it is receiving data from a student and subsequently providing an updated record for that student or taking a request from a lecturer for a list and providing it at short notice or processing an offer made by an admissions tutor so that the candidate hears from PCAS as quickly as possible.

The Registry is the heart of the student management system, maintaining the flow of data to and from a continuously changing spectrum of offices, individuals, committees, management, external bodies, etc. Here the central student record is maintained; here, the majority of the data that will eventually be used by all those groups and individuals is recorded. As well as student-related data, details on courses, on modules and their many separate sets of timetables, on tutors, LEAs and employers, on examinations and assessments, on medical certificates received, and on accommodation, are all gathered and processed.

Besides owning the data, Registry staff have considerable control over the operation of the student management system. They can choose whether to enter data themselves or to use the data preparation service in the computer centre; to enter data using real-time screen-based record editors or to use a monolithic batch mode multifile update system. They can choose when, and how often, to run updates.

Standard programs in the various user menus cover the functional needs of most areas of work of the office. For example:

- Many programs assist office staff in keeping relatively minor but essential files up-to-date covering data on personal tutors, courses, modules, fields, A level subjects, etc. Other programs provide updating facilities for many main files, either for groups of records or for individual records selectively presented on screens.
- Applications are processed, predominantly on the basis of tape exchange with PCAS, and visits and interviews are arranged, using subsystems to record the details, to print checklists for the many interviewers involved, and to print personalized labels for visitor/interviewee information packs.

- Examination results and awards made are recorded, and annual reports prepared for local education authorities.
- Enrolment forms are preprinted for most students before enrolment each autumn. These, together with marksheets and LEA reports, are produced on a turnround basis; specific computer files are created at the point of printing the forms so that subsequent data entry can be reduced to a minimum of keystrokes.
- From all of the data in the system, many different lists, reports, labels and statistics are prepared, on both a routine and an *ad hoc* basis.
- There are a great number of output programs available through a conveniently small number of menus. Each program contains a dialogue offering a wide range of choices affecting, for example, selection with respect to most data items, sort orders, print items per line, print lines per record, paging, number of copies, type of stationery, destination printer, real-time or batch execution, blackness of print image, redirection of printout to a computer file, creation of reference files to allow the same set of selected records to be used by another program, saving and naming selections for future use, local and remote printing, etc.

In addition, there are housekeeping options in the master menu such as inspection of the print queue and the batch queue; inspection of the files relating to the automatic recovery procedure; options to delete unwanted print files from past program runs (different programs retain different numbers of past print files); tasks associated with the data recovery system, access security and terminal log files; and control of the disc copies of PCAS tapes received.

For experienced student management system users there is a rich opportunity to develop personal computing. A third generation relational database management system, INFO, can be used for interactive file enquiries, and for writing, storing and running programs.

All enquiries, whether *ad hoc* or by standard program, can be made of the accumulated file of past students just as easily as of the file of current students.

Registry users of the student management system have come to feel secure in their data processing and take the assistance of information technology very much for granted. Their confidence in the system derives from consultation over proposed innovations, from training and from the degree of control over their work which has been consciously left with them. If the system has succeeded it is by enhancing that control.

Timetabling of students into their module classes is a task involving students, teachers and office staff. It requires considerable software development and computing resources. This requirement is greater if the processing involved includes, for example, deciding who will teach a class, when the class will be taught, and to what room the class will be assigned. The requirement will be smaller if the only decision is assigning each student to a set (variously called a group, laboratory class, split, etc.) for which the timetabled hours and rooms are already determined, and for which the teacher will be (or be provided by) the module leader.

The latter approach is the one used on the Modular Course. In practice, the sequence of events is as follows:

- The Registry prepares and circulates predicted numbers of students on next year's modules.
- A representative from each subject (module grouping) collates room requirements (overhead projectors, etc.) and set requirements (number of sets and preferred times, although there are rules guiding alternative set times to ensure even use of the teaching week, and balanced numbers in all sets of all modules).
- The Registry, through liaison with all such representatives, records these details together with notes on which departmentally owned rooms (e.g. laboratories) have already been booked and which poolable (centrally owned) rooms would be preferred.
- Rooms are then booked as part of a coordinated exercise covering both modular and non-modular courses. The Registry enters the booking data into the timetabling system.
- Before each term the Registry runs a timetable creation process. This takes the module programme of each student in turn, finds the best set for each module, and assigns the student to those sets. 'Best set' means several things, especially: absence of clashing hours, and balanced numbers in the sets for each module.
- The final stage is to print the students' timetables. This process simply involves, for each student, assembling the details of hours, days and rooms for each assigned set for each module, and printing it on a grid on a timetable form.

This approach has been popular with teaching staff who are not known for being slow to complain about their administrative load. The distinctive feature of the method lies in what it does not do. Lecturers retain control over their destiny. Each module leader determines which other colleagues, if any, will be involved in teaching the module, determines the times of alternative sets (within certain rules), and which departmental rooms (laboratories, etc.) will be used (by agreement with the Head of Department). Staff can also indicate preferences for any pooled rooms required (e.g. lecture and seminar rooms). The pitfalls of allowing a computer to attempt these tasks stem in part from their inherent variability. For example, a lecturer may have irregular commitments outside the Course, or the institution, which are not easy to record in a computer (unless complicated data structures are used) and are even harder to gather accurate information about in the first place.

There are several further ways in which the database of module timetable data can be exploited. For example, disabled students can be timetabled into ground floor rooms. It is also possible, though this is relatively undeveloped at present, to answer questions such as 'If this laboratory class is moved from Wednesday to Thursday, will the module have more balanced sets?'

There are many problems associated with the use of teaching accommodation that are probably not unique to Oxford Polytechnic: for example, bunching of teaching in the middle of the week with under-use of rooms on Monday and Friday, and rooms not being used at the booked times. Analysis of the cause of these problems indicates that solutions lie in changes in human organization and so no attempts have been made yet to automate the allocation of rooms to classes. It is notable, in this context, that the Modular Course timetabling system (including both the 'slotting' framework, and the timetabling process described above) favours no part of the week.

Departmental and faculty offices make limited use of the student management system. Some have requested and been given VDU access with tailored menus, subject to appropriate training and security arrangements being made. Occasional department-specific data items have been incorporated. The Modular Course involves all departments, and most students take modules in more than one department. Departments, being responsible for the subjects in which modules are taught, have information needs at the subject committee level. This involves, for example, production of detailed print outs for scrutiny at examinations time, including student records with progress messages, module performance analysis and medical certificate summaries. Fields, in which modules are grouped to form the basis of student programmes, need information to support the work of the Field Chairs. This will involve various lists and statistics for field reviews and field administration, for example, project registration, field trip arrangements, and module evaluation.

Departments and faculties receive staffing calculated from the student FTEs they teach. This is a fluid quantity that can slip from a department's grasp as students switch module registrations. The resourcing exercise involves the preparation and circulation of baseline figures at the census date early in each session, showing student FTEs by subject. These are calculated from the number of students registered for each module together with its size, single or double, rather than from class hours.

The *Library* relies on data from the student management system to create borrower file records within its automated circulation control system. Reports are subsequently produced for library staff to use in updating its student borrower records, for example, to enable a close watch to be kept on student withdrawals to encourage the return of books. Other name and course order lists are found useful in general library administration such as processing external readership applications for Polytechnic students.

The student identity number used by the Registry is the basis of the Library's borrower number. Bar coded identity numbers are printed by the registry and stuck on that part of the preprinted enrolment form which is detached at enrolment each year. This part is then laminated, with a photograph, to become a multi-purpose ID card (for use in the library; presented when collecting LEA award cheques; deposited against software loans at computer centre reception; displayed on candidates' desks during examinations, used to activate overnight access locks, etc.).

Predictions of student numbers on modules in future terms and years assist the Polytechnic bookshop in matching bookstocks to future demands.

Student Services is a user and a provider of lodgings and accommodation data about all students, including those on the Modular Course, and has access to the student management system for this purpose. The system also allows membership records of former students who have joined the Oxford Polytechnic Association to be maintained. Staff also use a software model, which embodies hall admissions policy and elements of wardens' experience, to draft lists of candidates to whom hall places should be offered.

Examinations on the Modular Course are termly and involve the registry, the examinations office and the various examinations committees in considerable use of student management system data processing. The marksheet for each module is returned to the examinations office by the module's internal examiner. It is checked and passed to the Registry who, together with the data preparation service, enter and verify approximately 10,000 individual marks and grades within a 36-hour period each term. For registry staff to monitor effectively the accuracy and completeness of this operation requires detailed record-keeping and effortless access to data. All of this the system provides.

The next stage is the printing of records for students in each of the various groupings for which formal scrutiny is carried out. The sheer quantity of calculation and printing required for this stretches the modest administrative processor to its limit for two days.

The MEC itself receives papers only for those students whose progress requires attention. However, for any other students whose case is raised a VDU immediately provides the Committee with the necessary details. The credibility of the system has brought streamlining and economy to this task. There has been no need to supplement the Chair's VDU with printers or copiers or to provide expensive video projection or multiple large-screen monitors to allow the whole Committee to view such data.

A detailed analysis of the performance of all students taking each module each term is prepared for the MEC. It is also distributed for appropriate action to subject committees.

Every individual student-module result since 1981 (currently over 120,000) is held in the student management system database in a form suitable for analysis for Course review and evaluation purposes. This has been exploited by the Course Evaluation Officer and other researchers and has informed debate on proposed structural changes to the Course.

The *Finance Office* is able to initiate tuition fee invoicing and to maintain a ledger of fees due and received, using its own student management system menus. Interactive access to individual student records is also provided, together with a subset of output programs tailored to the work of the office.

Course management, including the Dean, Assistant Dean and Course Co-ordinator, all have VDU access to those student management system options that can assist their work. Individual student data are available to assist counselling. Registry staff prepare the FTE analyses that are an essential part of staffing decisions, dialogue with NAB/PCFC, and admissions management. The system also provides the Course Evaluation Officer with assistance in sampling and in questionnaire distribution.

The *directorate* has a continual need for processed information to support decision making, to develop policy and advice for Academic Board and Governors, and to satisfy the appetite of external bodies such as DES, CNAA, NAB/PCFC, etc. for numerical information. From time to time there is also the need for high quality information to support particular initiatives, such as the (successful) application for accreditation. Some progress has been made in recording data on the Polytechnic's rooms and buildings, and on the teaching space requirements of individual courses. These data are analysed, along with the database on course details and students' course and module registrations, using the methodology in DES Design Note 44.

Ad hoc reports are an essential part of any system that aspires not merely to process data routinely but to permit management information to be creamed off the top. In public sector higher education there is also a steady stream of requirements for special lists or figures. Such needs would be met in a new system written today through the use of fourth generation languages and their user-friendly report generators. But established systems written in third generation COBOL, such as the student management system, must rely on other means.

For student management system users there is a choice. Any office, individual or organization can request special reports from the Registry who will either prepare the reports themselves or will seek the assistance of computer centre staff. Experienced users have direct access to the main data files using the INFO relational database management system query language. Its ability to link files of data to each other exploits the indexed-sequential structure of the system very effectively. For more challenging reports, or ones requiring specialist programming skills, unusual access rights, etc. requests are channelled through a registry representative to the information systems team (currently four FTE) in the computer centre. This team carries out information-related analysis and programming for all teaching and non-teaching departments.

Ease of preparation of *ad hoc* reports is exploited by the Registry in the explicit interests of the students themselves. A protocol agreed with the students' union provides the structure for permitting disclosure of student-based data. This allows the Registry to meet requests from student societies for lists and labels in support of membership activities, and to respond to requests from project students for assistance with questionnaire sampling and distribution. Help has been given to outside organizations such as professional bodies wishing to publicize associate membership to undergraduates, to foreign consulates with scholarships on offer to nationals, etc.

Coping with change. The transition of the Course to a new structure and regulations in 1985–86 presented a unique data processing challenge. The size of the advanced module credit was increased by 20 per cent. Various conditions for achieving each of the Course's many awards were changed. There were over 1,000 'transition' students, who had achieved advanced credits under the old course, who would have to be assimilated to the new course.

Different formulae for this assimilation were proposed and compared using a computer model of the old and new regulations. Particular formulae were chosen and a number of transition 'tariffs' were computed, covering thousands of eventualities. These were then incorporated into the relevant parts of the student management system software. Each transition student received, on each term's record, a personalized statement of exactly how the tariffs applied to his/her individual circumstances (see Figure 1.2). The progress summary of the student's record could then be compared directly with that statement. This was particularly important to tutors involved in academic counselling, and to the many staff involved in scrutiny of student programmes after each term's examinations.

In spite of its complexity, transition passed by leaving few students, teachers or office staff feeling perplexed. This might have been due to the support from the student management system. Or perhaps the transition adventure was just a ripple on the ocean of complexity in which all newcomers to the Course must quickly learn to sink or swim. Or both.

There are some registry subsystems that contain key features of expert systems. For example, the COMPOSE model automatically assembles knowledge about an applicant's first-year module preferences, his or her chosen field, 'safety net' field and entry qualifications, about which modules are available, their prerequisites and their timetable slots. The rules about compulsory modules and alternative routes for fields individually and in combination, together with rules embodying the experience of admissions tutors, specific teaching constraints, and academically desirable aspects of module programmes in some subjects, are then applied to the assembled data. The best possible module programme, out of hundreds of thousands of possible programmes, is then determined and recorded for that applicant. At the same time, the model ensures that runs of the same module in different terms receive balanced numbers of students.

Security issues, covering data protection, data integrity, security of access, physical security, recovery arrangements, and printout control, have been given close attention. Periodic security audits involving key users and computer centre staff are carried out, and documentation and training implications are considered. Besides passwords (for both users and menus) a range of devices are incorporated including disk configuration, direc-

Table 7.1 Oxford Polytechnic student management system: data flow and users' operational overview

Data comes from:	Students Staff PCAS LEAs Examiners Committees, etc.
on media such as:	Forms Letters Voice Magnetic tape Turnround documents Turnround computer files Electronic mail
It is recorded by:	Departmental staff Registry staff Data preparation
using:	Local VDUs Screen-based forms
and running:	Realtime editors Tape control menus Monolithic updates
It is stored in indexed-sequential files relating to:	Students, annotations, module programmes, medical certificates; fields, courses; modules, sets; tutors; employers, LEAs; candidates, choices, conditions, comments; interviewees, visitors; entry qualifications; accommodation; hall applicants; fee transactions
It is retrieved by:	Individual record display *Ad hoc* enquiry Standard output General selection
in various modes:	Batch Online
Outputs are in the form of:	Lists Tables Turnround documents Special forms Labels Computer files
for example:	Stock listing Marksheets Student records Bar codes Letter quality A4 Enrolment forms Timetables Address labels Index cards Transcripts Name labels Magnetic tape User directories (for personal computing) Floppy disks System directories (for turnround document processing) System directories (for input to other programs)
They can be printed:	Locally, in remote departments, in the Computer Centre
on various printers:	Daisy wheel, dot matrix, laser, line printers
with options such as:	Multiple copies, extra black image, plain side of listing
They are sent to:	Staff, students, Oxfordshire CC offices, departments, faculties, course management, directorate, Field Chairs, personal tutors, senior tutors, library, student services, researchers, publishers, LEAs, CNAA, DES, NAB, PCAS
by:	Hand, Electronic mail, Internal post, Collected, Student pigeonholes, Post Office

tory-based access control, physical keys, data switch programming, VDU access control and other 'proprietary' features.

Log files are kept for every log-in session for every student management system user. These record all screen and keyboard activity and provide the basis both for programme debugging in the event of the unexpected happening, and for determining whether a user needs advice or further training. Further journal files are kept of all file activity permitting recovery of virtually all data in the event of catastrophic disc failures or other disasters of comparable severity, using 'roll-forward' protocols. These files are also designed to assist software audit in the event of system failures.

The *Computer Centre* has carried out the programming work for all of the systems described above. The project management and systems design has been carried out using staff based within the Modular Course. These staff have also reported to the Chief Administrative Officer, which has ensured seamless integration between modular and non-modular data processing. The designers have liaised very closely with computer centre staff. The liaison has been with all of the functional groups within the computer centre: analysis and programming, operations, user services, communications and technical support and data preparation. The success of the student management system, and the many subsystems that now pervade the management and administration of the Polytechnic, is built on these close links. Although the links may evolve (for example, to mesh in better with the systems developments and other preparations for corporate status) the close cooperation between users and computer centre staff continues. The value of this relationship shows itself when users need support in emergencies, or when large new projects have to be undertaken. The closeness of consumer and producer has made it easier for the Registry to expect and to receive from the Computer Centre a computing service of commercial standards, especially in terms of quality and meeting deadlines.

As in all sections of higher education, resources are severely constrained and the question of balance between support for teaching and non-teaching departments has to be answered. In practice, the balance has varied, in response to changing teaching and non-teaching needs. Management must be ever watchful to ensure that the balance is never unreasonable.

Scarce programming resources can be swamped by maintenance work on existing systems, to the detriment of work on new developments. One effective solution to this has been to identify elements of maintenance that can safely be put into the hands of the users. For example, patrolling disk space, access supervision, tape file control, and session log file administration can all be packaged for safe and effective control by suitably trained administrative users. They will relish the additional control this gives them over their work. 'Virtuoso' users can play a particularly useful role in minimizing calls on support programmers, and in training other users.

The communications infrastructure required by the student management system is also provided by the computer centre: all of the terminals and printers are in the same cross-campus network as the many hundreds of terminals for academic use (in open access terminal rooms, staff offices, laboratories, etc.). A very high (and carefully monitored) level of security is none the less maintained. Being part of this broader user community brings benefits. Use of electronic mail is escalating and supplements the work of the student management system, for example, as a means of transmission of reports, and of requests for information. Other general purpose applications software, initially provided to meet teaching needs (such as word processing, and spreadsheets) is also available to student management system terminal owners. Any terminal in the network may also be used to book one of the Polytechnic cars, or squash courts.

The *future* for the student management system is clear. The present pattern of add-on development will continue. Projects currently under consideration include cohort analysis, BTEC course administration, text storage and retrieval, improved responsiveness of the timetabling system to student needs, allocation of resources (e.g. rooms to departments, or staff to courses), and examination timetabling. In addition, links will be developed to the new financial systems that have been installed for the independent Polytechnic. This will permit, for example, the integration of student-based data into course-costing exercises. Routine maintenance and the extension of the system to more departmental users will continue to take up time, as will dealing with the planned 40 per cent growth in the FTEs on the Modular Course between 1986–87 and 1991–92.

Summary

People, as the users of the student management system, are the most important part of it. Control of their own work or study must be enhanced, rather than diminished, by the system. The need for student-related systems to cope with growth and change is met by designs incorporating flexibility in data structures, office procedures, software options, and software support. The system mirrors the life of the Modular Course and of the institution. It enables change without dictating it.

Technical note

Student management system programs are written mainly in COBOL77 and run on a Prime 2755 minicomputer rated at 1.6 mips with 4 Mb of main memory and 1.1 Gigabytes of disk storage. The 2755 is a traditional uniprocessor running the proprietary Primos operating system. Up to 48 concurrent asynchronous lines are available for printers and VDUs. These lines pass through the main Polytechnic data switch, a Micom 2000, which permits approximately 100 end-users to connect to the 2755 (up to 48 simultaneously). All VDU lines run at 9600 baud. The VDUs themselves are all of Televideo 925 standard or higher; many now have the higher TVI 955 functionality. The end-users include approximately 60 office automation workstations based on Nibus AT clones with access to the Prime, e.g. for archiving purposes and for email. The SMS programs are controlled by macros written in PLP (a subset of PL/1) and CPL (Command Processing Language). These macros are initiated through menus tailored to individual users. Approximately 20 main data files, and 15 subsidiary files, are organized as indexed-sequential with most files having several secondary keys. Local or 'slave' printers are supported and are common for users outside the Registry. The Registry itself has two addressable network printers: a 400 cps data processing dot matrix, and a lower letter quality thimble printer. Central laser printing is used on a rapidly growing scale. The local office lines are grouped together at time division multiplexors connected to the data switch by twisted pairs. Personal productivity aids for users are mainly network-based and include the INFO rdbms from Doric, S2020 spreadsheet from Access Technologies and Composer word processing from Wordmarc. These three products have a limited level of integration. There is also locally written email which is heavily used and includes a full JANET interface. Local personal computing based on AT micros, probably Research Machines Nimbus AX workstations, is planned for 1988–89. Computer Centre staff manage SMS development using Project Manager Workbench on a Nimbus AX20.

8

Afterword: 'Going Modular'

David Watson

Much of the interest expressed in the Oxford Polytechnic Modular Course by visitors and correspondents is on behalf of departments or whole institutions who are considering 'going modular'. Several have established research programmes or 'task forces' charged with explaining to their Academic Boards the costs and benefits of such a move. These highly programmatic intentions pose a dilemma. The Oxford Polytechnic scheme is *sui generis*: the extent to which it has developed in a particular context of constraints and opportunities makes many of the decisions it has made and many of the systems it has adopted difficult to transfer to other institutional frameworks. Among the peculiarities of the Oxford scheme which other institutions find hard to replicate are the following:

- *The size of the module*
  By sticking to a small basic unit (100 or 120 hours of student effort) delivered across terms (three per year, and with the development of a summer school, potentially four) Oxford has established a pattern of regular assessment, and regular award-making examinations boards, that depends upon an intensity of academic staff commitment that other institutions would find difficult if not impossible to implement within the current framework of practice and conditions of service.
- *The cost and complexity of the student management system*
  All of the administrative systems developed on behalf of the modular course have been designed and tested in-house, beginning with a simple program for pre-enrolment information and student progress on a microcomputer in 1972 and leading up to the large and sophisticated management system described in Chapter 7. To make an equivalent investment in one slice, rather than over the 15-year span recorded in this volume, would be an enormous burden. Oxford's experience in making a major shift of data-processing infrastructure in 1980–81 gives us some 'feel' for what would be involved.

- *Institutional ethos*
  The success of the Modular Course over the past 15 years has in many respects been the success of Oxford Polytechnic. The Course is the most distinctive and, on this scale, the most innovative contribution that the Polytechnic has made to the public sector in higher education. Commitment to it, and its future, is high at central, departmental and individual levels. The largely voluntary history of its development, set out in Chapter 2, attests to the way in which many courses within the Polytechnic have seen their interests as lying within the modular scheme, and the way in which new subject areas and approaches have then influenced it from inside. Much of the Polytechnic's negotiation with outside bodies, such as the funding agencies referred to in Chapter 3, reflect this corporate commitment. It is hard to imagine another institution of comparable size achieving this kind of ethos as a result of a single administrative decision, however consensual its support.

Broadly, there have been two ways of 'going modular', each with several internal dimensions and variables.

- *The 'creation' model*
  This implies the *ab initio* design of a new course, including (as in the Oxford case) the dissolution or suspension of an old course. Such a course can then remain relatively isolated, or at least on a par with other courses offered within the same institution on a more traditional basis, or act as a focus for further course development (through the processes of 'accretion' or 'invention' discussed in Chapter 2).
- *The 'conversion' model*
  This implies either an internal (course-led) or external (faculty- or institution-led) decision to recast and develop current courses in a modular form. Commitment to the development often varies in relation to the extent of internal or external steering of such a change. There have been several instances of polytechnic or college academic boards adopting a 'modular policy' which requires specified courses, over time, to move into a common framework of delivery (usually achieved administratively by common timetabling and academically by convergence of course titles and examination arrangements).

  Some conversion experiences have been more successful than others. One pitfall has been the temptation to adopt a 'lowest common denominator' approach to unit length and course delivery, leading to a set of central 'modular' regulations which, in their efforts to retain local features of individual cases, present the student with an overall scheme of mind-boggling complexity. There are several such 'phantom' modular courses around, which in their presentation to applicants and to enrolled students of a unified title ('Critical studies', 'Combined studies', etc.) imply promises (of internal transfer, for example) which they then cannot deliver.

Other key points within the conversion strategy include the implications for the course or department of 'opting in' (what controls, for example, are relinquished over admissions, timetabling, examining, or future resource planning?) and the significance for contributing departments of continuing 'non-modular' work. One possibility, usually disastrous for the development of the wider course in terms of resource constraints, is the 'header-tank' model, where departments with strong freestanding courses resource and develop these first before turning to the faculty- or polytechnic-wide scheme.

Departmental or institutional motivation for making these decisions is, of course, a crucial consideration and inspires its own typology. Roger Waterhouse of Wolverhampton Polytechnic (on behalf of the Midlands Consortium for Credit Accumulation and Transfer) has effectively put this point in historical perspective. In introducing a workshop on 'modularity and credit accumulation' in 1986 he identified two broad phases:

Modular Developments: Phase One
The idea for this workshop came out of a realisation that Modularity was very much back on the agenda in a wide variety of institutions. I say 'back on the agenda', since some of us were engaged in designing Modular systems back in the late '60s and early '70s. That was the period of expansion – of new ideas and new ventures – when arguments about flexibility and student choice, new subject combinations, new structures and inter-disciplinarity were very much to the fore. This was the phase which saw the development of the big modular schemes at City of London and Oxford Polytechnics. It saw the development of a Modular Humanities Degree then DipHE, at Middlesex. It saw large Modular Combined Studies degrees at Manchester, Sunderland, Central London and other Polytechnics. It also saw at least two complete institutions – Hatfield Polytechnic and Crewe and Alsager College – adopt modularity as an organising principle for (almost) all their work.

Modular Developments: Phase Two
The second phase of modular developments belongs very much to the '80s. It is the era of retrenchment, cuts and post-NAB planning. In the changed climate, the most powerful arguments surrounding modularity have to do with resources, rationalisation, how to maintain options with declining rolls, and ultimately – sheer survival. An important difference is that modular courses are now not being planned from scratch – they are being made by mergers and amalgamations of pre-existing courses. This is the era in which the Middlesex modular scheme was put together out of four pre-existing courses – two modular and two non-modular. When the Lancashire Combined Studies scheme was made by putting together two Combined Studies courses. When the NELP Multi-subject DipHE was

> hacked out of bits of many pre-existing courses. When Bradford College put together its 6 AFE courses into a single scheme. And if now a whole institution, as Wolverhampton Polytechnic has done and Essex Institute well might, resolves to modularise its whole course provision, it starts with a portfolio of courses which looks virtually complete.

In the remainder of his paper Waterhouse's intention is to urge preservation of some of the radical and progressive emphases of the pioneers into the new instrumentalist era. His themes – student control, variable entry, the ladder of intermediate awards, exemption, prior learning and learning contracts – are all still attainable if the 'conversion' experience of modularity moves beyond the formal redefinition of courses to the genuine 'culture of negotiation' implied by the best courses. What follows is a check list of issues to be considered in preparing for and making the decision. Its overall theme is the importance of linking motives with implementation, of reflecting the answers to the 'why?' questions in the 'how'?

Modularization: strategic issues

Why?	*Notes*
Academic development	How does it fit into academic plans of departments/faculties, the institution?
Features claimed include:	
Choice	Flexibility for students
Interdisciplinarity	A context for curriculum development
Relevance	e.g. 'employability' record
Expanded opportunities	Access/'mixed mode'/intermediate awards
Economics	What are the precise forecasts in terms of resources, including hidden costs?
Features claimed include:	
Scale	Common teaching, large groups
Retention	Opportunities to change/readjust programmes
Flexibility	e.g. in the deployment of staff
Protection of unpopular subjects	The 'seamless web' argument
How?	
Scope	
Issues include:	
Inclusion/exclusion	Basis for selecting or rejecting courses

Voluntary/prescriptive	Are participants invited or compelled?
Relationship to other schemes	Credit transfer consortia, BTEC professional bodies, etc.
Planning and development	
Issues include:	
Purposes and extent of consultation	
Level of policy formulation	Who decides? e.g. department, faculty or Academic Board
Advice	limits of external advice
Structure	
Decisions required about:	
Units	Variety allowed, how defined (e.g. in terms of student effort)
Length and awards	
Regulations	Including examination committee structure
Implementation	
Strategies available between:	
Preliminary investigation	Pilot schemes
Phased introduction	
'Big bang'	Need for transitional arrangements
Management	
Decisions required on:	
Centralization/decentralization	Of specific functions
Countervailing interests	e.g. departmental priorities
Evaluation	Loops back to the why questions; have you achieved what you intended/anything else?

This set of essays has sketched the development, from small but imaginative beginnings, of a major Modular Course with significant institutional support. The question of 'going modular' does recur even now at Oxford Polytechnic, but on the local basis of whether or not to apply to join a going concern. As for redefinition or reorientation of the overall enterprise, this material should serve to confirm that such developments are difficult but still possible. Oxford Polytechnic is not (yet) an entirely modular institution.

Appendix I Slotting

Chris Coghill

The Modular Course teaching and examination timetables are based on a slotting system, the essential elements of which are *slots* and *rules* for the use of slots. A slot is a pair of hours. There are 21 slots in the formal timetable. For example, slot 1 is the two hours 9.00–10.00 a.m. and 11.00–12.00 noon. The slots and their hours are defined as follows:

	Monday	*Tuesday*	*Wednesday*	*Thursday*	*Friday*
9–10	1	5	9	11	15
10–11	2	6	10	12	16
11–12	1	5	9	11	15
12–1	2	6	10	12	16
2–3	17	3		7	13
3–4	18	4		8	14
4–5	17	3		7	13
5–6	18	4		8	14
7–9	21	19		20	

Some use may be made of other hours but 1.00–2.00 p.m. and 6.00–7.00 p.m. are recognized as meal breaks and teaching is not normally arranged at these times. In addition, at Oxford Polytechnic Wednesday afternoon is not used for teaching on full-time or sandwich courses in order that student society activities may be organized. Since the Modular Course is agnostic as to mode for teaching purposes (though not for other purposes such as progression and fee assessment) no modules have formal timetabling on Wednesday afternoons.

Slots are allocated to each module in order to cover lectures, practicals, seminars, problem classes, etc. Tutorials and other more occasional needs are arranged outside the formal timetable, for example, a lecturer would normally arrange tutorial times directly with the students on his or her module after the beginning of the teaching term.

Modules requiring only one or two formally timetabled contact hours per week would have only one slot allocated. There are few such modules. Much more common is the need for three or four weekly contact hours, for example, in arts and social studies modules. Such a module would have a pair of slots allocated. These pairs are not random. They are normally the pairs (1+2, 3+4, etc.) that provide four-hour blocks within a morning or an afternoon.

The majority of science modules, together with the more practical-based modules in other subject areas (for example, some modules in education, geography, catering, visual studies, etc.) require five or six hours of formal timetabling. Frequently, this is made up of three hours for a practical and two or three hours of lectures. To meet this need specific groups of three slots are used as follows:

Basic modules (i.e. those normally taken in Stage I)	Advanced modules (i.e. those normally taken in Stage II)
1+2+(3 or 4)	(1 or 2)+3+4
5+6+(7 or 8)	(5 or 6)+7+8
9+10+(11 or 12)	9+10+(11 or 12)
11+12+(13 or 14)	(11 or 12)+13+14
15+16+(17 or 18)	(15 or 16)+17+18

There is a symmetry in this convention which has the following important result: for basic modules, a block of three hours is available in the morning; advanced modules have a three-hour block in the afternoon. A student in Stage I, taking only basic modules, will have all of his or her practicals in the mornings. This will ensure that no afternoon lectures for one module conflict or clash with the practicals for another module. Similarly, all advanced modules have their practicals in the afternoons.

The value of this arrangement is that where a module in a student's programme has alternative practical classes (i.e. splits, sets or groups) in order to cope with large numbers of enrolled students, it is possible to timetable the student into more than one of those sets. This in turn enables balanced numbers of students to be achieved in the different sets of a module.

The asymmetry of the Wednesday slots which cover only the morning is turned to advantage. There are some instances of advanced modules which, although they can only be taken in Stage II in the relevant single fields, are compulsory or recommended in Stage II in cognate double fields. To avoid restricting the timetable for students taking such modules they are usually assigned slots 9 and 10, i.e. Wednesday morning. This avoids conflict between lectures and practicals for the Stage I students on such modules.

Teaching needs have shaped the current slotting timetable. Specific needs have required exceptional treatment. For example, core language studies require contact hours and laboratory work to be spread through the week. This is achieved by grouping such studies into compulsory modules, notionally assigning slots to each module according to the standard pattern for modules requiring six weekly contact hours, and then shuffling those slots between the modules so that the hours for each module are spread out horizontally across three days rather than vertically and mainly within one day.

How are slots allocated to a module? When a new module is first proposed it is a requirement of the course review process that the proposal includes draft slots for the module. These are worked out by the field concerned in conjunction with the Course Co-ordinator. They will take into account the existing slotting for related modules in order to maximize the opportunities for students in relevant fields to take it.

Once slots have been allocated to a module it is not easy to change them since the module will have been incorporated into student programmes on the basis that its existing slots do not clash with those of the other modules in those programmes. However, there is an established procedure for considering proposed slot changes in order to meet the changing needs of fields. Such proposals for basic modules must normally be made by 1 December, 22 months before the academic year in which the change is to occur. Proposals for advanced modules must normally be made by 1 December, before the academic year in which the change is to occur. A change of term is equivalent to a change of slot. In emergencies, slot changes can be approved within these deadlines. In all cases, Field Chairs must take any necessary steps, including scrutiny of individual module programmes, in conjunction with the course administration, to ensure the viability of any student's module programme that might be affected.

Most modules enrol large enough numbers of students to require more than one group to be timetabled for activities other than lectures. While all students may attend the same lecture hours, they may be split into several groups for seminars, practicals, etc. Since students compose their module programmes on the basis of the slots assigned to each module, and since these further seminar or practical groups will involve different slots, it is essential that the timetable for each module should include at least one set (the first set) where the only hours used are within the essential or 'advertised' slots for that module.

For the subsequent or further sets of a module, the seminar or laboratory times are chosen by stepping forward through the week at the same time of day as in the first set. If this time is in the afternoon then Thursday follows Tuesday since there is no teaching on Wednesday afternoons. This rule, applied to all modules, tends to increase the number of groups that a student can attend for each of his or her modules. This in turn allows the timetabling system to spread the students across the groups with each module in order to achieve balanced numbers within the groups. This permits accommodation to be used efficiently and is essential for planning the teaching, technical support, etc. for a module.

Where a module uses three of the four hours in a block it is usually the last three of those four. This allows staff to prepare laboratories in the hour beforehand. It is also important (where there is more than one set for a module) that no lectures are given within those three hours since this would hinder the flexibility for students to be timetabled into alternative blocks for any of their modules.

Appendix II
Modular Course Regulations

Contents

Transition Regulations

Students who entered Stage II before September 1985 are subject to separate transition regulations issued to them as leaflet M13, available from the Registry.

Glossary of Terms

An explanation of many of the terms used in these regulations is contained in the Glossary published in the *Modular Course Handbook*.

BA, BEd and BSc Degree and Honours Degree, DipHE, and Certificate Regulations

1. Registration

(i) *Certificate*
A candidate for a Certificate must register for two single fields or a double field.

(ii) *DipHE*
A candidate for a DipHE must register for two single fields approved for the award of DipHE, one of which may be the Applied Education field, or for an approved double field.

(iii) *BA and BSc*
A candidate for a BA or BSc degree or honours degree must register for two single fields or for a double field.

(iv) *BEd*
A candidate for a BEd degree or honours degree must register for *either*
(a) the Applied Education single field and one other approved single field, and must also register for the age range to which his or her professional training will be directed, *or*
(b) the BEd for Qualified Teachers.

2. Stage I

(i) (a) A candidate for BEd for Qualified Teachers must pass the module '*Contemporary Education Issues*', within two years. All other candidates must satisfy (b) and (c) below.
(b) A Stage I programme shall include a minimum of *ten* module credits available in Stage I, including those compulsory for the student's two single or one double field(s). Students are advised to take *twelve* module credits in total.
(c) To complete Stage I a student must pass, within a period of two years, at least *ten* basic module credits including the compulsory modules for his or her field(s), and must satisfy any specific requirements (see 16 below) of his or her field(s).

(ii) A student taking the *Catering double field* is required to complete satisfactorily one year's industrial placement in order to complete Stage I.

(iii) A student taking one or two of the single *language fields* is required to complete satisfactorily a period abroad (normally one academic year) in order to complete Stage I.

3. Stage II

(i) Before entering Stage II a student must complete Stage I or be granted exemption from it.

(ii) A candidate for an award must satisfy the appropriate requirements set out in paragraphs 6, 7, 8, and 9 below.

(iii) A student who claims exemption from part of the course may not count in the further module credits required for a degree, honours degree or diploma any module from which further exemption can be claimed by means of equivalent qualifications or prior study.

(iv) A limited number of basic module credits may be included in the further module credits required for an award. The limitation is as follows:

no more than two (for a degree or honours degree candidate) module credits other than any which are acceptable in stage II; no more than four (for a degree or honours degree candidate) or three (for a diploma candidate) basic module credits overall.

Candidates for the award of the BEd for Qualified Teachers are limited to one basic module credit.

Such basic module credits must not be those from which exemption can be claimed by means of equivalent qualifications or prior study.

4. Pace of Study

The student will be required to pass at least *three* module credits a year in order to continue with the Course. He or she will normally be required to pass at least *seven* module credits a year in order to continue with the Course as a full-time student and in order satisfactorily to complete a year of full-time study.

5. Fee Assessment

For the purpose of assessing the tuition fees payable a student will be classified as full-time or part-time. To be classified as part-time in any academic year (or in any group of three consecutive terms) the student must not take more than *six* module credits in that year (or in those three terms).

A student whose complete programme of modules extends for only one or two terms will not be classified as part-time if the total of module credits taken averages more than two per term.

6. Certificate

To obtain a certificate a student must complete Stage I except that

(i) a student taking the Catering double field is not required to complete one year's industrial placement, *and*

(ii) a student taking one or two of the single Language fields is not required to complete a period abroad.

7. Diploma of Higher Education

To obtain a diploma a student shall have qualified to enter Stage II of the course and within three further years shall have:

(i) fulfilled the specific requirements for his or her field(s);

(ii) passed at least *eight* further module credits including at least *seven* acceptable for a double field, a single field, or two single fields.

8. Degree Requirements

(i) *Registration*

Any student wishing to be a candidate for a degree without honours must register this fact before or during the term in which he or she expects to complete the field and course requirements.

(ii) *Field and Course Requirements*

To obtain a degree without honours a student shall have fulfilled the specific requirements for his or her field(s) and shall have fulfilled the requirements of sub-paragraph (iii) or (iv) below.

(iii) *BA and BSc Degree*

To obtain a BA or BSc degree without honours a student shall have qualified to enter Stage II of the course and within five further years shall have passed at least *sixteen* further module credits including at least *fourteen* acceptable for a double field or at least *fourteen* acceptable for two single fields including at least *seven* for each field.

(iv) *BEd Degree*
To obtain a BEd degree without honours a student shall have qualified to enter Stage II of the course and *either*
(a) within *seven* further years shall have passed at least *twenty-five* further module credits including at least *eighteen* acceptable for the Applied Education field and at least *seven* acceptable for another approved field, *or*
(b) within *four* further years shall have passed at least *eleven* further module credits including *nine* module credits acceptable for the BEd for Qualified Teachers.

9. Honours Degree Requirements and Classification

(i) *Registration*
A student is assumed to be a candidate for an award with honours unless he or she registers for another award.

(ii) *Field and Course Requirements*
To obtain an honours degree a student shall have fulfilled the specific requirements for his or her field(s) and shall have fulfilled the requirements of sub-paragraph (iii) and of either sub-paragraph (iv) or (v) below.

(iii) *Project, Dissertation and Synoptic Module Requirements*
To obtain an honours degree the student shall normally have passed *either* a double *or* two single project or dissertation modules *or* a synoptic module.

(iv) *BA, and BSc Honours Degree*
To obtain a BA or BSc honours degree a student shall have qualified to enter Stage II of the course and within five further years shall have passed at least *eighteen* further module credits including at least *sixteen* acceptable for a double field or at least *sixteen* acceptable for two single fields including at least *seven* for each field.
The class of the honours degree shall be decided on the average of the *eighteen* further module credits in which the candidate has been awarded the highest marks in accordance with the schedule at sub-paragraph (vi) (a) below.

(v) *BEd Honours Degree*
To obtain a BEd honours degree a student shall have qualified to enter Stage II of the course and *either*
(a) within seven further years shall have passed at least *twenty-seven* further module credits including at least *eighteen* acceptable for the Applied Education field and at least *seven* acceptable for another approved field, *or*
(b) within five further years shall have passed at least *sixteen* further module credits including *fourteen* module credits acceptable for the BEd for Qualified Teachers.

The class of the honours degree shall be decided on the average of *either*
(c) in the case of a candidate registered for the Applied Education field, the *twenty-five* further module credits in which the candidate has been awarded the highest marks in accordance with the schedule at sub-paragraph (vi) (a) below, *or*
(d) in the case of a candidate registered for the BEd for Qualified Teachers, the *sixteen* further module credits in which the candidate has been awarded the highest marks in accordance with the schedule at sub-paragraph (vi) (a) below.

(vi) *Classification*

(a) *Schedule*

Schedule	*Minimum*
First Class Honours	70%
Upper Second Class Honours	60%
Lower Second Class Honours	50%
Third Class Honours	40%

(b) *Synoptic Examinations*
For the purposes of honours classification the mark for a synoptic examination will count as two module marks and may be substituted for one or two inferior marks (if any) arising from those modules on which classification is to be decided in the corresponding field.

(vii) *Limitation on modules taken*
A candidate for an honours degree specified in *column A* below completing Stage II in a period longer than that specified in *column B* shall not be permitted to take more than the number of futher module credits specified in *column C*:

A	*B*	*C*
BA, BSc	*6* terms	*21* credits
BEd for Qualified Teachers	*Any* period	*20* credits
Other BEd	*9* terms	*32* credits

10. Medical and other Evidence

(i) The Examination Committee shall have power to award marks after consideration of course work and any other appropriate evidence of attainment to a candidate who, through causes outside his or her control attested by appropriate written evidence such as a medical certificate accepted by the Committee, has failed to complete modules or to sit a synoptic examination.

Such a candidate will normally be awarded a mark by the Committee after consideration of relevant course work and subsequent or concurrent results.

(ii) The Examination Committee may allow an extension of a student's course by an equivalent time if, through causes outside his or her control attested by written evidence such as a medical certificate accepted by the Committee, he or she missed a term's work or more.

(iii) The Examination Committee shall have discretion to award an *aegrotat* diploma or degree to a student who, for medical reasons attested by written evidence accepted by the Committee, has missed a term's work or more and who, for any reason, is unable to accept an extension to his or her course.

11. Projects and Dissertations

(i) A project or dissertation module may be a single or a double credit module.
(ii) For a double field student a double project or dissertation module in that field will count wholly towards that field.
(iii) For a student taking two single fields the project or dissertation will whenever appropriate be inter-disciplinary and in that case one module credit will count towards each of the fields involved.

(iv) If a double project or dissertation is set wholly within a single field, only one of the two credits will count toward the minimum required for that field but both will count towards the total number of acceptable credits required.

(v) All credits for project or dissertation modules will count towards the student's overall total of further module credits required for the award of a diploma, degree or honours degree.

12. Advanced Entry and Credit Transfer

(i) A student granted partial exemption from Stage I of the Course by means of exemption from less than eight module credits shall be required to meet the requirements for the completion of Stage I as specified by the Modular Admissions Committee.

(ii) A student granted exemption from Stage I of the Course by means of exemption from *eight* to *twelve* module credits shall be required to meet the normal conditions in Stage II to qualify for an award.

(iii) A student granted partial exemption from Stage II as a result of prior qualifications (including the DipHE) and/or experience shall meet requirements set out in the schedule in the *Appendix* to these Regulations.

(iv) A candidate for honours granted partial exemption from Stage II who is registered for a field for which a pass in a synoptic examination is normally required must pass such an examination. Sub-paragraph 9(vi)(b) above applies.

(v) A student participating in an approved exchange arrangement with another institution may count up to *four* credits towards requirements for a degree or degree with honours. The classification of the degree with honours shall be based on module credits earned at Oxford Polytechnic in accordance with the schedule referred to in 12(iii) above.

(vi) A student admitted with exemption must, in order to gain an award, *either*:

(a) complete successfully a minimum of one year's full-time study, *or*:

(b) obtain at least half the number of credits normally required for an award, when that number of credits may be obtained in less than one year.

13. Grades

(i) As an indication of student progress and, in the case of pass grades, for inclusion in the final transcript the Examinations Committee shall award grades for modules and synoptic examinations according to the following scheme:

Grade		*Percentage Mark*
A	pass grades	70–100
B+	pass grades	60–69
B	pass grades	50–59
C		40–49
S	Pass for modules with pass/fail assessment only	
R	Entitled to re-assessment without retaking module	0–39
F	Not entitled to re-assessment without retaking module	0–39
P	Pass at re-assessment	40 (maximum awardable)

FR	Fail at re-assessment (where the marks at the initial assessment and at re-assessment differ the higher mark will be awarded).	0–39

(ii) A student awarded an *R* grade shall be entitled to re-assessment at the end of the following term or in the vacation prior to the next academic session. Re-assessment implies either re-examination or the submission of further course work for evaluation or a combination of the two.

(iii) No more than one re-assessment shall be allowed in any term or in the vacation prior to the academic session, and a student awarded *R* grades for two or more modules may normally choose in which module to be re-assessed.

(iv) At a re-assessment only grades *P (40%)* or *FR* shall be awarded.

(v) A student who has failed to pass a *School Experience* module may, at the discretion of the Examination Committee, be permitted an extension of one further term to take the module again. This discretion will be exercised on not more than one occasion for each student.

14. *Transcript*

Upon completion of his or her studies or upon the award of a *DipHE* or *Certificate* the student will receive a transcript stating (a) the award, if any, made, (b) the modules passed and grades obtained, and (c) in the case of BA, BEd and BSc awards the field or fields studied.

15. *Degree Titles*

(i) *BA and BSc Candidates Registered for Double Fields*

The title of the degree awarded to a successful candidate will be:
Bachelor of Arts or *Bachelor of Science* in (*double field name*)

(ii) *BA and BSc Candidates Registered for Single Fields*

The title of the degree awarded to a successful candidate will be:
Bachelor of Arts or *Bachelor of Science in Combined Studies* (*name of first field*) and (*name of second field*).

(iii) *The Determination of Bachelor of Arts and Bachelor of Science*

For the purpose of determining the award as BA or BSc each field is classified as *Arts (A)*, *Science (S)* or *Neutral (N)* and the following rules will apply:

A=*BA*	S=*BSc*
A+A=*BA*	S+S=*BSc*
A+N=*BA*	S+N=*BSc*
N+N=*BA* or *BSc*	*to be approved individually by Academic Board*
A+S=*BA* or *BSc*	

(iv) *BEd Candidates*

The title of the degree awarded to a successful candidate will be:

(a) for a candidate registered for the Applied Education field:
Bachelor of Education

(b) for a candidate registered for the BEd for Qualified Teachers:
Bachelor of Education for Qualified Teachers.

16. *Specific Field Requirements*
All references in these Regulations to specific field requirements are to the approved requirements of individual single or double fields. These are printed annually in the *Course Handbook*.

17. *Interpretation*
In cases of dispute these regulations shall be interpreted by the Academic Board in conformity with the *CNAA 'Principles and Regulations for the Award of First Degrees and Diplomas' (1979)*.

Regulations for Associate Students and for the Diploma in Advanced Study

1. Associate Students

Associate students are those studying a module or a selection of modules for personal enrichment or for the purpose of qualifying for a higher level course. Mature students without formal qualifications or lacking general entry requirements may be advised to take a selection of basic modules in order to qualify for entry to the Modular Course. Such programmes of study for associate students do not lead directly to an award on the Modular Course.

2. Selection of Modules

An associate student may register for modules only with the approval of the Course Co-ordinator. Students must normally possess the module prerequisites although these may be waived at the discretion of the module leader.

A student's selection of modules will not be divided into Stage I and Stage II.

Both *basic* and *advanced* modules are available to associate students.

3. Restrictions

If a limit to the number of students on a module has to be imposed then students registered for an award will take precedence over associate students.

Failure in *three* successive or concurrent modules will normally preclude associate students from registration for any subsequent modules.

4. Polytechnic Diploma in Advanced Study

This Diploma can be awarded by the Academic Board on the recommendation of the Modular Examinations Committee to a student registered for the award who satisfactorily completes a programme of *seven* or *more* module credits including at least *six* advanced module credits. This programme must be approved by the Field Chairs for two single fields or Field Chair for one double field or single field. The student must complete the requirements for the Diploma within eight terms.

Candidates for the Diploma may not simultaneously register for a CNAA award.

Candidates for the Diploma are subject to Regulation 4 (*Pace of Study*) and 5 (*Fee Assessment*) on page 141 above.

5. Medical and other evidence

The Examination Committee shall have power to award marks after consideration of course work and any other appropriate evidence of attainment to a candidate who, through causes outside his or her control attested by appropriate written evidence such as a medical certificate accepted by the Committee, has failed to complete modules.

Such a candidate will normally be awarded a mark by the Committee after consideration of relevant course work and subsequent or concurrent results.

6. Grades

As an indication of student progress and, in the case of pass grades, for inclusion in the final transcript the Examinations Committee shall award grades for modules according to the scheme in Regulation 13 (*Grades*) sub-paragraphs (i)–(iv) on pages 144 & 145 above.

7. Transcript

Upon completion of his or her studies an associate student may apply to the Registry for a transcript stating the modules passed and grades obtained. A student awarded a *Diploma in Advanced Studies* will be issued with a Modular Course transcript in conjunction with the award of the Diploma.

References

Ball, R. and Halwachi, J. (1987). 'Performance indicators in higher education', *Higher Education*, 16, 4.

Bradley, C. (1984). 'Sex bias in the evaluation of students', *British Journal of Social Psychology*, 23.

Bristow, Steven *et al.* (1985). *Modular Course Working Party Report*, Wolverhampton Polytechnic.

CNAA Consultative Paper (1987). *Future Strategy: Principles and Operation.*

Lindsay, R. O. and Paton-Saltzberg, R. (1987). 'Resource changes and academic performance at an English polytechnic', *Studies in Higher Education*, 12, 2.

McBean, E. A. and Lennon, W. C. (1985). 'Effects of survey size on student ratings of teaching', *Higher Education*, 17, 2.

Miron, M. (1988) 'Student evaluation and instructors' self-evaluation of university instruction', *Higher Education*, 17, 2.

Moses, I. (1986), 'Self and student evaluation of academic staff', *Assessment and Evaluation in Higher Education*, 11, 1.

NAB circular of 29 March 1985 on 'Modular and combined studies courses'.

Oxford Polytechnic (1972, 1978 and 1983). *Application to the CNAA for approval and renewal of approval of the Modular Course.*

Parlett, M. and Hamilton, D. (1972). *Evaluation is illumination: A New Approach to the Study of Innovating Programmes*, Centre for Research in the Educational Sciences, University of Edinburgh, Occasional Paper, No. 9.

Theodossin, Ernest (1980). *The Modular Market*, FE Staff College, Coombe Lodge.

US Joint Committee on Standards for Educational Evaluation (1981). New York, McGraw Hill.

Waterhouse, Roger (1986). 'Modularity and credit accumulation' (copies available from the Midlands Consortium for Credit Accumulation and Transfer).

Watson, David (1983). 'A worm's eye view of mission control', *Times Higher Education Supplement*, 12 August.

Watson, David (1985). 'The Oxford Polytechnic Modular Course 1973–83: a case study', *Journal of Further and Higher Education*, 9, 1.

Index

The Society for Research into Higher Education

The Society exists both to encourage and coordinate research and development into all aspects of higher education, including academic, organizational and policy issues; and also to provide a forum for debate – verbal and printed.

The Society's income derives from subscriptions, book sales, conference fees, and grants. It receives no subsidies and is wholly independent. Its corporate members are institutions of higher education, research institutions and professional, industrial, and governmental bodies. Its individual members include teachers and researchers, administrators and students. Members are found in all parts of the world and the Society regards its international work as amongst its most important activities.

The Society discusses and comments on policy, organizes conferences, and encourages research. Under the imprint SRHE & OPEN UNIVERSITY PRESS, it is a specialist publisher of research, having some 40 titles in print. It also publishes *Studies in Higher Education* (three times a year) which is mainly concerned with academic issues; *Higher Education Quarterly* (formerly *Universities Quarterly*) mainly concerned with policy issues; *Abstracts* (three times a year); an *International Newsletter* (twice a year) and *SRHE News* (four times a year).

The Society's committees, study groups and branches are run by members (with help from a small secretariat at Guildford), and aim to provide a forum for discussion. The groups at present include a Teacher Education Study Group, a Staff Development Group, and a Continuing Education Group, each of which may have their own organization, subscriptions, or publications (e.g. the *Staff Development Newsletter*). A further *Questions of Quality* Group has organized a series of Anglo-American seminars in the USA and the UK.

The Governing Council, elected by members, comments on current issues; and discusses policies with leading figures, notably at its evening forums. The Society organizes seminars on current research, and is in touch with bodies in the UK such as the NAB, CVCP, UGC, CNAA and with sister-bodies overseas. It co-operates with the British Council on courses run in conjunction with its conferences.

The Society's conferences are often held jointly; and have considered 'Standards and Criteria' (1986, with Bulmershe College); 'Restructuring' (1987, with the City of Birmingham Polytechnic); 'Academic Freedom' (1988, with the University of

Surrey). In 1989, 'Access and Institutional Change' (with the Polytechnic of North London). In 1990, the topic will be 'Industry and Higher Education' (with the University of Surrey). In 1991, the topic will be 'Research in HE'. Other conferences have considered the DES 'Green Paper' (1985); 'HE After the Election' (1987) and 'After the Reform Act' (July 1988). An annual series on 'The First Year Experience' with the University of South Carolina and Teesside Polytechnic held two meetings in 1988 in Cambridge, and another in St Andrew's in July 1989. For some of the Society's conferences, special studies are commissioned in advance, as *Precedings*.

Members receive free of charge the Society's *Abstracts*, annual conference Proceedings (or *Precedings*), *SRHE News* and *International Newsletter*. They may buy SRHE & Open University Press books at discount, and *Higher Education Quarterly* on special terms. Corporate members also receive the Society's journal *Studies in Higher Education* free (individuals on special terms). Members may also obtain certain other journals at a discount, including the NFER *Register of Educational Research*. There is a substantial discount to members, and to staff of corporate members, on annual and some other conference fees.

Further Information: SRHE at the University, Guildford. GU2 5XH UK (0483) 39003